M000278591

Technology and the End of Authority

Jason Kuznicki

Technology and the End of Authority

What Is Government For?

Jason Kuznicki
Cato Institute
Washington, District of Columbia, USA

ISBN 978-3-319-48691-8   ISBN 978-3-319-48692-5   (eBook)
DOI 10.1007/978-3-319-48692-5

Library of Congress Control Number: 2016962628

Printed on acid-free paper

This Palgrave Macmillan imprint is published by Springer Nature
The registered company is Springer International Publishing AG
The registered company address is: Gewerbestrasse 11, 6330 Cham, Switzerland

When we have dropped the idea that the history of power will be our judge, when we have given up worrying whether or not history will justify us, then one day perhaps we may succeed in getting power under control.

—Karl Popper

Whoever had created humanity had left in a major design flaw. It was its tendency to bend at the knees.

—Terry Pratchett

ACKNOWLEDGMENTS

Among many others, I am indebted to Grant Babcock, Evan Banks, Trevor Burrus, Nicholas Geiser, Gene Healy, Jeremy Kolassa, Brink Lindsey, Jon Meyers, Darren Nah, Alex Nowrasteh, Marco Orsimarsi, John Paslaqua, Aaron Powell, Timothy Reuter, John Samples, Julian Sanchez, and Nick Zaiac of the Cato Institute. I am further indebted to Elaine Fan and Chris Robinson at Palgrave Macmillan. I am indebted to the Fellows at Liberty Fund collectively, and to Sarah Skwire and Amy Willis in particular. I am indebted to Richard Avramenko, Carrie-Ann Biondi, Jason Brennan, Eric Nelson, Victoria A. Russell, Curtis Tate, Derek Lee, Steve Horwitz, Jacob T. Levy, Patrick Cahalan, Alex Tabarrok, and to Timo Virkkala for the term "agnarchism." I owe particular thanks to Adam Gurri, whose feedback and support were exceptionally valuable. I thank several anonymous readers as well. I owe many discussions of political theory online to the *Ordinary Times* blog community. And finally, I am indebted to my husband, Scott Starin, whose help and support I am honored to accept.

Epigraph from Sir Karl Popper's *The Open Society and Its Enemies* (New York: Routledge Classics, 2003), by permission of the Karl Popper Copyright Office, University Library of Klagenfurt.

Section entitled "Kant: Government Is an Unfinished Project" adapted from *Arguments for Liberty*, Aaron Ross Powell, editor (Washington, DC: Cato Institute Press, forthcoming), used by permission.

CONTENTS

CHAPTER 1

Introduction: The March of God in the World

"The march of God in the world, that is what the state is," wrote G.W.F. Hegel in *The Philosophy of Right*. Blasphemous or not, I think many will find Hegel's claim to have some truth to it: The state is commonly thought central or focal to the rest of society, almost in the manner that God himself is to a believer.

Hegel's view has been a common one in the history of political philosophy. This book will dissent from it and present an overview of the history of western political philosophy that is openly skeptical of its subject matter.

By my way of thinking, the state is *not* the central institution of society. It is peripheral, or, at the minimum, it should be thought of that way as a necessary corrective to our existing, and bad, mental habits. The state should be thought of as a dumping ground, a place to finally, dejectedly put our society's problems when we have decided, rightly or wrongly, that we have no better way of solving them. Sometimes a society may do well by entrusting these things to the state; sometimes it may suffer. Sometimes the state deals with these cast-off problems rather effectively, and sometimes it does not.

In any case, however, the state should be understood as a last resort. We ought to be embarrassed rather than proud whenever we reach for the apparatus of government to solve a problem. The use of the state is always an admission that either our other social technologies have failed us or we have prematurely abandoned them.

What exactly do I mean by the state? I mean the thing that political philosophers have discussed over time. To be sure, there have been enormous

© The Author(s) 2017
J. Kuznicki, *Technology and the End of Authority*,
DOI 10.1007/978-3-319-48692-5_1

differences among different types of states, both in the real world and in philosophers' minds. Even in the ancient world, the Greek *polis* was radically different from the Persian Empire, and Greek political philosophers would probably never have accepted that the lessons meant for a *polis* could apply to an empire.

Ever since then, however, both philosophers and statesmen have been relatively willing to disregard these distinctions, and to talk about "the" state as if it were a single thing with a coherent intellectual history. I intend to take them at their word, at least provisionally, and see what may result.

On many different questions, historians are apt to divide into two rival camps: the lumpers and the splitters.[1] Lumpers like to draw big, bold pictures. They comment on the similarities and the connections, even at the expense of detail. Splitters, meanwhile, are all about the distinctions. Splitters love nuance, and they are meticulous where the lumpers are bold.

In this book, I will be a lumper. I will look at the thing that has been called the state, and I will also look at some of the other things—such as the ancient Greek *polis*—that have frequently been talked about as effectively synonymous. I know that this is not an unproblematic approach. But lumpers and splitters alike will do well to remember: there are no unproblematic approaches. Along the way, and particularly regarding ancient history, I will try to illustrate some of the strengths and also weaknesses of the move that I have made toward lumping. In part, we meet with a tradeoff: What we lose, for example, in specificity to the situation of ancient Greece, we gain insofar as later political philosophers took ancient Greece to be a model for their own times. To a great extent, political philosophers continue to do so today. As a result, one can hardly enter into the history of the discipline without a bit of lumping.

In criticizing the tendency to think of the state as the central or focal aspect of society, I will not claim that I have identified and held up for criticism a specific ideology or a consistent, trans-historical set of beliefs. Instead, I will look at some of the key thinkers in western political thought, identifying commonalities among them. These individuals were often at odds with one another about deep and important matters, but I will argue that despite their differences, they all shared this tendency to place the state at the center or the forefront of life, and to hold that the state was very closely aligned with the good as they understood it.

This ought to strike us as weird when we consider what the state actually does, which is to *govern*: The state appropriates from some, distributes to others, and issues commands and prohibitions. None of these things seem particularly central; each seems, rather, remedial in nature at best.

Yet aside from religion, no discipline outside of political philosophy arrogates to its subject matter such a place of social importance. It has only been with the emergence of modern anarchism in the nineteenth and twentieth centuries that the state's centrality has seriously come into question, and even then—even among the anarchists—one still finds a remarkable degree of state-centered thinking: The state is not the march of God in the world, but the march of an all too powerful devil. In relatively few cases it has been seen as a set of entirely human problems that are badly in need of solutions, perhaps by other, less violent, methods.

Thinking of the state as an entity tasked with using a flawed solution on a set of otherwise intractable problems may ultimately leave us with a better public discourse and—just possibly—a better state: Only fools and megalomaniacs suggest improving upon the march of God in the world. All by itself, this fact may explain a good deal of the history of the state. The state's purported centrality, its purported majesty, has distorted our thinking about it. A heap of problems, though, is something we can all maybe get to work on, temperately, reasonably, and with the aim of gradually solving or eliminating them through more refined methods than those of organized violence.

Yet what I will suggest in the concluding sections of this book is not a technocratic, problem-solving approach to government. Rather, I favor what might be termed technocratic de-government: The attempt to develop technologies, both physical and social, that will allow us more often to do without the coercion that is inherent to governing and thus to the state. We have made great strides in this direction already, even without having done so intentionally, and I will discuss some of the ways that technology can be understood as an ongoing challenge to political authority.

In time, this process may even leave us with nothing we might call a state at all. This is a frightening prospect for many. Whether it will come to pass, I do not know, and I take what I term an *agnarchist* view of political philosophy. Quite simply, I have found myself surprised to discover that I can't come down on one side or the other in the seemingly endless philosophical debates about whether utopia happens to have a state or not. I never have been to utopia, and if you haven't either, then I will commend to you the virtues of agnarchism.

The question of whether we need a state may still have a provisional answer, however, and I think that here the answer is yes: We do still require a state, but only because it does not yet appear that we know how to do better. Those who would do without a state probably need to theorize much less, and experiment much more.

The plan of the book will be as follows.

Part I will review some of the classic texts and thinkers of political philosophy, paying particular attention to their use of rhetoric and to the knowledge claims they make, implicitly or explicitly, regarding what the state is competent to do and what it is supposed to be doing. I begin in the ancient world. Why is it, I ask, that so many authors have given the state so much to do? Did states commonly (or ever) succeed at it? Did they even try?

What emerges is a history of grand, all-encompassing dreams. For the most part, it is also a history of crashing failures, or of systems that never left—and could never leave—the printed page, chiefly for want of technology. I note as well the emergence of a minority opinion—namely that the state is something other than the be-all and end-all of society. The voices on this side are fewer and less influential, to be sure, but they are clear, and in earnest, and I believe that they have been the more correct. I will examine some of the reasons for the ascendancy of what I consider the less correct view; this is a theme I will revisit as well in Part II.

Ordering Part I will be a typology of answers given to our focal question—*What is government for?* And again, I will be a lumper. I find that the dominant view was as given above: The state was a central, transcendent, or all-encompassing element of society; governing was of course its business. Closely aligned to this view, and often scarcely distinguishable from it, was the view that the state served as a venue for the exercise of individual virtue, particularly of individual martial virtue, and that this exercise of virtue entailed individual self-sacrifice to the needs of the state. Dissenters existed, as I have mentioned, and these commonly described the state as a curse, an affliction, or simply as a harmful social convention. A remarkable few of them suggested that the state had a real but limited role to play. The dissenters, however, were in most eras a minority, and a much less esteemed one at that.

My selection of texts and thinkers in Part I is guided by an attempt at presenting a representative sample from various major schools of political theory. As a result, I have had to make some difficult choices: Why Bossuet, for example, and not Filmer? Why the relatively obscure Oppenheimer and not Spencer (or Hayek!)? And so on.

In many cases, these were not easy choices. Some authors, like Tommaso Campanella, are obscure or even uninfluential in the mainstream history of political theory. Yet I would argue that Campanella both illustrates the enduring appeal of utopianism and in his own day served to propagate that tradition. Few have read Campanella himself, but it hardly matters; all have read his successors.

Other choices are perhaps entirely too familiar. Readers who don't want to take in yet another summary of Locke or Hobbes should feel free to skip these sections, or indeed anything in Part I that feels too pedestrian for them.

Part of what guided my selection, besides an attempt at representation, has been that each author or school will be useful in some way to the argument that follows in Part II. Although I am primarily writing a history, I am also writing an argument about history. In Part II, I will argue that two key technological/economic developments are bringing about the demise of the all-powerful central state: First, history is increasingly being written not just by the governing classes, but by many different types of people, owing to the rise of bourgeois wealth and inexpensive authoring and publishing technologies. This allows new perspectives to emerge, not merely those that serve the central state. And second, I will argue that the ability to instantiate real-world attempts at utopia—also owing to technological advances—has prompted a reckoning. Powerful states may be beautiful when put into words, but we have found them horrifying in practice.

I will also present the outlines of a different way to think about the state: The state is *not* central to society; its work, governing, is what we do when other options have either failed or been neglected. We can understand, however, why so many people wrote that the state was central, and we can begin to reconstruct our thinking along other lines if we so choose. The fields of history and economics, particularly in the twentieth century, have done a great deal to problematize the state as the purported center of human society, and political philosophy could learn a great deal from them. Arguably it's already doing so.

The errors of the past have almost all lay in the direction of too much state action, not too little. We should therefore suspect that we may still be making similar errors. There is every reason to believe that many are still undiscovered, and that many seemingly solid reasons for state action are not solid at all. Intellectuals would appear to have been systematically too quick to invent reasons to trust the state, and too slow to invent ways of doing without it. It is up to us to correct this bias, and we have no particularly good reason to think that the work is finished. Although its precise outlines cannot be traced, a society that does not place the state at its center may stand to realize more fully the kind of freedom to which we humans are entitled.

NOTE

1. Charles Darwin first coined these terms to describe rival approaches
   to species classification in biology, but they are applicable in many
   other fields as well. In the discipline history, their use can be traced
   to J.H. Hexter, "The Burden of Proof." See also Immanuel Kant,
   Critique of Pure Reason, p. 526 ff., Ernst Cassirer, *The Myth of the
   State*, p. 6.

PART I

A Too-Brief History of Political Theory

CHAPTER 2

The Ancient State and the Myth of Marathon

Following the battle of Marathon in 490 BCE, a messenger went to Athens to share the news of the Greek victory over the Persians. He ran the entire distance, which is sometimes said to be 42.195 kilometers, or 26 miles, 385 yards. Completely exhausted, he uttered the words "Joy to you, we've won," and then he died.

Like many ancient stories, the episode is shrouded in confusion. The Greek historian Herodotus was the first to mention anything like it, but he placed the heroic run *before* the battle. He claimed that it was not from Marathon to Athens, but from Athens to Sparta—a much longer distance. The runner, named Pheidippides, did not die at the end; he remained very much alive and asked the Spartans for help, which they eventually granted.

Five hundred years later, the Roman historian Plutarch appears to have gotten confused. He wrote that the runner's name was Eucles, that the run followed the battle, and that its course was from Marathon to Athens (actual distance: closer to 40 km). "Hail! We are victorious!" Eucles supposedly said, and then he died.[1]

Another century passed, and the essayist and satirist Lucian wrote of another Athenian runner, this time named Philippides—a name with a familiar ring to it. This Philippides, however, ran from Marathon to Athens, breathed his message of victory, and died, just as Eucles had done in Plutarch's account. (Actual distance between Marathon and Athens: still about 40 km.)[2]

The story took root in the ancient imagination and, later, in the modern imagination.[3] For the Romans, the first marathon was the story of

© The Author(s) 2017
J. Kuznicki, *Technology and the End of Authority*,
DOI 10.1007/978-3-319-48692-5_2

a life well-spent: It had been spent in the service of the state, forsaking all other causes in the process. Was the run necessary? Was it tactically advantageous? These seem to be secondary considerations. Pheidippides' run from Athens to Sparta was both longer and very clearly advantageous, because it secured the help of the fearsome Spartans. But Plutarch—or whoever first altered the story—appears to have thought that he was improving it by making the courier die for a bit of information, news that would surely have arrived sooner or later anyway.

The myth of Marathon illustrates how ancient historians and many who have followed after them have viewed the relationship between the state and the individual: Individuals are judged as successful or failed, as heroic or base, only insofar as their lives touch on the interests of the state. Man is not the measure; the state is. In a sense, that is precisely what government is *for*: By demanding self-sacrifice for the needs of the community, it allows the individual to manifest the essence of human virtue.

It was for a similar reason that Herodotus made Solon, the lawgiver of Athens, declare, in a story of doubtful authenticity, that an Athenian named Tellus had been the happiest of men. We know little about Tellus, but Herodotus had Solon describe his happiness as follows: Tellus lived in a city that was prosperous; he had had sons and grandsons; and he was materially comfortable. And yet he was not to be counted truly happy until he had died fighting victoriously for his country: "He fought for his countrymen," wrote Herodotus, "routed the enemy, and died like a brave man; and the Athenians paid him the high honour of a public funeral on the spot where he fell."[4]

No one could be counted fully happy, Solon claimed in this story, until he too had had such a death. The story of Tellus also illustrates what the ancient historians took to have been their own purpose in writing: History prepared men for service to the state, telling them what service was expected of them and identifying the good with obligation to the state. As Polybius wrote, the study of history is "in the truest sense an education, and a training for political life." Those who do not care for history are "idle" and "indifferent."[5] And Livy wrote that one reads history to

behold instances of every variety of conduct displayed on a conspicuous monument; that from thence you may select for yourself and for your country that which you may imitate; thence note what is shameful in the undertaking, and shameful in the result, which you may avoid.[6]

Another work by Plutarch, the *Parallel Lives,* further illustrates the point. The *Parallel Lives* was a series of biographies of prominent Greeks and Romans. Yet ultimately it did not aim to provide insight into *individual* lives. On the contrary, the *Lives* amounts to a course in how to serve the state, in whatever role, high or low, that fate may place you.

One cultivated one's character, and one read Plutarch, so as to better serve the state. As with the first marathon, many of the *Lives* ended in a heroic death in the state's service. These included the *Lives* of the Greeks Cimon, Lysander, Nicias, Pelopidas, Philopoemen, and Pyrrhus. They also included Lycurgus of Sparta, whose death deserves special mention for creativity: Before traveling to the oracle at Delphi, Lycurgus forced the Spartans to swear to abide by his laws until he returned. While abroad, Lycurgus deliberately starved himself to death, thus making his city's laws eternal.

Relatively few Romans had similar, self-sacrificing deaths. Marcellus, who commanded during the Second Punic War, is perhaps the most notable example. It is unclear to me whether Plutarch meant it as a reproach that so many eminent Romans died in their beds of sickness or old age, but he might have. Whatever the case, and without exception, every individual chronicled in the *Lives* helped to shape or to serve the state, to either good ends or bad ones. Their stories were selected precisely to illustrate such actions.

It is understandable, then, that Plutarch devoted much space to Alexander the Great. In his lengthy biography, one story stands out in particular.

One day a trader brought King Philip of Macedon a horse named Bucephalus. Plutarch writes that Bucephalus was "savage and intractable"; Philip refused Bucephalus, thinking the horse was wild. But Alexander replied: "What a horse they are losing, because, for lack of skill and courage, they cannot manage him!"

As horse is to rider, so the ruler is to his subjects. Good government serves to master everyone, even a foolish and unruly people. Much of the rest of Plutarch's *Life* of Alexander shows him quelling or resolving conflicts among his subjects and courtiers. The portraits we get of his underlings can be taken as commentaries on the proper place of individuals in relation to government: Subjects are to be obedient and to place the state's interests ahead of their own; those who do not are justly punished.[7] For example:

Philotas ... had a high position among the Macedonians ... However, he displayed a pride of spirit, an abundance of wealth, and a care of the person and mode of life which were too offensive for a private man ... [A]s a young man will often talk freely in vaunting and martial strain to his mistress and in his cups, he used to tell her that the greatest achievements were performed by himself and his father, and would call Alexander a stripling ... [Alexander], on hearing the story, ordered her to continue her meetings with Philotas and to come and report to him whatever she learned from her lover.[8]

We moderns are not apt to regard the story so highly. To us, it may seem little more than court gossip, and we have no use at all for a manual detailing how to be the proper subject of an absolute monarch. Indeed, we might prefer that Plutarch had written on virtually any other topic to which he might have had access.

What else might he have written about? Let us contrast the story of Philotas to the Antikythera mechanism, a remarkably complex analog computer constructed in the first or second century BCE, probably by craftsmen working in the tradition of Archimedes. The mechanism could calculate the dates of festivals, the positions of heavenly bodies, and the eclipses of the sun and the moon. And yet we know the device was real only because we moderns found it in a shipwreck, not because any ancient author ever bothered to describe its use, its invention, or even its existence.[9]

Yet to present-day sensibilities, an ancient treatise describing the construction and use of the Antikythera mechanism would be a priceless historical document. And the greater dissemination of engineering knowledge might even have been of humanitarian benefit to Rome itself: Slaves performed much of the construction, transportation, and other hard labor that went into making the city of Rome so pleasant for the upper classes. The slaves' lives might conceivably have been eased, and slavery itself might even have been ended, had the Romans valued the invention of labor-saving devices over the enslavement of additional labor. One modern scholar has even concluded that first-century CE Roman engineering possessed all the elements necessary to build a working, piston-driven steam engine.[10]

The Empire's failure, of both engineering and morals, is a sort of miniature of my overall argument: Historically, societies have resorted to the state, and statesmen have called for state action, after two things happen. First, a problem is defined; and second, compulsion is thought the proper solution to that problem. Manuals of imperial government were in a sense *substitutes* for manuals of engineering; the one taught how to win and keep slaves, while the other might have taught how to have a likewise comfortable life without slaves.

In a similar vein, consider the Archimedes Palimpsest. This text was a compilation of several works by Archimedes and others, in which the great mathematician solved many problems in geometry and pointed the way toward solving many others. One work contained therein, "The Method of Mechanical Theorems," contained the first use of infinitesimals in mathematics.

It would have been only a short step from Archimedes' method to integral calculus. But the ancient Greeks never took that step, and the only known copy of "The Method of Mechanical Theorems" was erased in 1229, so that a Christian liturgical text could be written over it. Only in 2008 was the text finally reconstructed using sophisticated multispectral imaging and computer analysis. By then Archimedes' work was no longer a stepping stone to higher math. It was a historical curiosity, and a sad one at that.

It would be a mistake to put all the blame on the specific monk who performed this one erasure. But his act was representative of a larger tendency: How many times must someone have copied the unreliable histories of Cassius Dio, or made yet another copy of Plutarch, rather than preserving this mathematical heritage, which might have been saved? What gifted early medieval scholar might have read and profited from "The Method of Mechanical Theorems," but could not find it? Might he have developed integral calculus, a thousand years before Newton and Leibniz? And there he sat in a library full of manuals for enslaving an empire.

Professional historians usually avoid counterfactual questions like these. They cannot be answered with any great rigor, and they usually shed little light on the history that actually happened. But it seems clear that in the history *of* history, much was foregone, or abandoned, for what amounts to a chronicle of evil.

It is painful even to contemplate where the arts and sciences might have ended up if scribes and copyists had picked anything other than imperial statecraft as their principal, and yet essentially barren, subject. We are still ignorant about so many other topics in ancient history, and perhaps we always will be: How exactly were the pyramids built? What did ancient music, particularly biblical music, really sound like? What was Greek fire made of? Marcus Terentius Varro (116–27 BCE) may have been the most learned man in Rome; he wrote dozens of books, but only one survives intact—and in it, he describes, unmistakably, *a germ theory of disease.*[11] What else might he have imagined? What might we have had today, if these foundations had been built upon, rather than lost?

Notwithstanding our more diverse interests, rulers and their reigns dominate ancient history. To the ancients, the state was the reason for history itself. This mode of writing undoubtedly influenced much of subsequent social thought, including philosophy and statesmanship, down to the present day. The state has been all too central ever since, and other concerns have been pushed to the margins.

Ancient philosophers mostly agreed with the historians' view or even took it further. Many of the philosophers' prescriptions gave more power to the state than the state actually enjoyed in any society that they knew. And some of their theories called for an order of pervasive state power that even modern technology cannot supply.

Plato was undoubtedly the first important political philosopher in the western tradition, and it is remarkable the degree to which his *Statesman, Republic,* and *Laws* all assigned a place of absolute importance and absolute authority to the state. The state was to touch everything in life; the state was to regulate all. Whether by written law or otherwise, the statesman was to be in charge of the citizens' lives, and citizens were to live for the purposes, and by the rules, laid down by the all-wise rulers. A good state could be expected to tell you what to do about virtually everything, and you were to obey it in all things.

To Plato, the best conceivable form of government was to have on hand at all times a *statesman.* This was no ordinary diplomat or legislator; for Plato, the term had a special meaning. It denoted a man who was so wise that he could rule case-by-case on every possible matter in the conduct of a city. The Platonic dialogue known as the *Statesman* is a description, and a defense, of the rule of such a man.

A sufficiently wise statesman would decide all controversies individually, in whatever way he thought best: "The political ideal is not full authority for laws but rather full authority for a man who understands the art of kingship and has kingly ability," Plato wrote.[12] When a man is truly wise, we should not let anything stand in his way, and any checks on his power are always going to be illegitimate. His own wisdom already checks any bad actions he might take, and ours cannot do better. Certainly to him a constitution, which is only a set of things that we non-statesmen have imperfectly scribbled down, should never serve as a barrier.

Alas, true statesmen are rare. They are so rare, Plato lamented, that in the whole world there might only be "one or two" at a time, if there were any at all (*Statesman* 293a). Yet letting ordinary folk govern themselves would be as foolish as letting them practice medicine on themselves. To prevent government by the less worthy, statesmen wrote down the laws. Once they did, all others were to follow the laws with unquestioning obedience. The statesmen themselves were of course free to transgress when the situation demanded it (*Statesman* 300–301c). They alone would know when it was necessary.

In Plato's view, government served to enact the will of a very small number of exceptionally wise people. We owe a particular obedience to the laws—what philosophers call *political obligation*—because we cannot expect to improve on the efforts of statesmen. This account also served to explain the existence of written law itself.

Plato left two dialogues that might testify to his own skill as a statesman, the *Republic* and the *Laws*. The *Republic* may be the most widely read piece of secular western philosophy, while the *Laws* is somewhat less often studied. Yet it is not entirely clear to me why this should be so. The *Republic* is frequently enigmatic and equivocal, while the *Laws* is entirely clear as to its intentions and meaning. Still, it is worth looking at them both in turn, beginning with the *Republic*, which Plato is known to have written first.

In the *Republic*, Socrates and his companions addressed the true meaning of justice. Determining the nature of justice as regards individuals turned out to be exceptionally difficult for them, and, in the early part of the dialogue, they arrived at no fully satisfying answers. They had similar trouble in fixing the relationship between happiness and justice. As an attempt to crack these two difficult problems, Socrates proposed examining the justice to be found in an ideal city instead: "If we found some larger thing that contained justice and viewed it there, we should more easily discover its nature in the individual man," Socrates claimed (*Republic* 434e).

Both individual souls and cities, Socrates claimed, consisted of three parts: reason, spirit, and appetite. Justice in each consisted of all parts being kept to their proper place. In a city, the three parts were constituted by philosophers, corresponding to reason; lovers of honor, corresponding to the spirit; and those who work for material or monetary gain, corresponding to the appetite. Each had its proper place, and the laws of the *Republic* aimed at keeping all of them in it: Those who work for material or monetary gain are to provision the city with material goods. Those who crave honor are to defend the city in battle. The philosophers would maintain the laws, attend to the harmony of the city, and seek to bring it ever closer to the ideal.

Much here conflicts with the *Statesman*. The philosophers of the *Republic* cannot possibly be statesmen; if they were, their city would have no need of laws. Nor was their city crafted by a statesman, for if it were, then everyone should have to follow the written laws without any deviations. And yet the philosophers were tasked with crafting and improving the laws, actions that the later *Statesman* forbids ordinary men. Plato appears not yet to have reached some key conclusions of the *Statesman*, although his authoritarianism is already clear.

The upper two classes, the philosophers and the guardians, are to maintain a community of possessions for nearly all forms of property. They are even to do the same with wives and children. At one point, Socrates mentions it in passing—as if it were obvious to the wise—that his republic rests on "other principles that we now pass over, as that the possession of wives and marriage, and the procreation of children and all that sort of thing should be made as far as possible the proverbial goods of friends that are all in common" (*Republic*, 423e–424a). It is one of the most remarkable asides in all of philosophy, given the scope of what it proposes.

The *Republic* does describe the community of wives, which amounts not to open marriage or free love, but to a strict calculation of the ideal childbearing partnerships, a surprising and heretofore unknown technology. It sounds to the novice like good deal of unfathomable numerology (*Republic* 546–47). The relevant portion runs as follows:

> Now for divine begettings there is a period comprehended by a perfect number, and for mortal by the first in which augmentations dominating and dominated when they have attained to three distances and four limits of the assimilating and the dissimilating, the waxing and the waning, render all things conversable and commensurable with one another, whereof a basal four thirds wedded to the pempad yields two harmonies at the third augmentation, the one the product of equal factors taken one hundred times, the other of equal length one way but oblong—one dimension of a hundred numbers determined by the rational diameters of the other dimension of a hundred cubes of the triad ...[13]

The *Republic* offers in effect one of history's first proposals for eugenics, and the above is its purported rationale. Its arrival here is deeply embarrassing to the eugenic enterprise in any era at all: Whereas modern eugenicists have promised what they term a scientific eugenics, the appearance of eugenics in the *Republic* shows that people are well-disposed to like the idea of eugenics with or without actual evidence in its favor. And it is evident, of course, that a government will be necessary to implement the plan. Impulses that lead us to find unevidenced claims attractive should be severely mistrusted, whether they bear the name of science or not, and particularly when they lead to strong claims of authority over others.

Plato's numerology should not be written off in favor of the more appealing aspects of his work. On the contrary, the numerology is all but inseparable from the rest. Without it, there is no separation of orders, who are born according to numerology. There are, therefore, no social harmonies and no philosopher kings, whose special knowledge is the program of divine begettings. The numerology is the origin of the entire society; it is the pillar that holds up the whole theory of government in the *Republic*. The three orders of the polity are not to undergo any improper commingling, and it is numerology that asserts their essential difference. Plato's discovery of the allegedly perfect human breeding number "is the basis of his historicist sociology," wrote Karl Popper, and I am inclined to agree.[14]

And yet attentive students of the *Republic* may recall, correctly, that the entire system of orders is also declared to have been founded on a *noble lie*. There was no natural basis whatsoever for it; nor was there any divine institution. The ruling class undertook these divisions solely for the preservation of its own power. It was ascribed to the gods, but it was undeniably of human origin.

Plato himself could not possibly have believed in the cornerstone of his own system; rather, he found it instrumental to *claim* that he believed it, and to encourage others to believe it, even though he knew it to be false. This presents a real difficulty for Platonists, because without the divine number, and without governing populations according to the dictates of a numerical system, the work of statecraft would seem to bear little resemblance to the world of Forms. Without numerology, politics cannot participate in the eternal, and it might not even be a proper subject of philosophy.

Plato seems to have earnestly desired that there *should* exist a discoverable body of knowledge, one knowable beyond all possible doubt, that would instruct rulers in the governance of the *polis*. One would govern by applying it to the lives of others, such that they produced purebred citizens of the three orders of the polity. And yet this technology does not exist, and the *Republic* therefore constitutes the first of a type we will often encounter: a polity that might work, if only the technology existed to make it so.

It would probably seem to us more likely that no field whatsoever, not even history or economics, was competent to produce a body of knowledge that allowed for a state to exercise perfect control over society. Yet a body of knowledge of exactly this type is absolutely necessary for a total state to operate successfully. What is worse, the body of knowledge must be unfailingly true if the total state is to succeed. This much Plato saw clearly. The lack of infallible knowledge, though, is a fundamental problem of all total states. Without it, total states produce lies, and they call them noble. They can hardly do otherwise.

It is hard to overstate the spirit of comprehensiveness that animates the *Republic*. The community of property in nonhuman goods, as practiced by the guardian class, encompassed nearly everything, and again, the state controlled it all: "[N]one must possess any private property save the indispensable … [N]one must have any habitation or treasure house which is not open for all to enter at will. Their food, in such quantities as are needful for athletes of war sober and brave, they must receive as an agreed stipend from the other citizens … so measured that there shall be neither superfluity at the end of the year nor any lack" (*Republic* 416d–e).

It is undoubtedly appealing to imagine such measures, and to imagine them succeeding. Yet emotional appeal, whether to readers or to authors, can serve only to make an idea propagate and endure; it cannot make the idea true or successful. One of the themes of this book is that canonical political theory, with Plato at its head, is beset by a bias in favor of things that are enjoyable to think about. And it is simply more pleasurable to design systems, and to imagine them working perfectly, than it is to imagine us all muddling through somehow.

The polity of the *Republic* will, of course, regulate education; each of the three classes is to be taught from earliest childhood that they are distinct from the rest through a divine plan. As we have already mentioned, this is plainly false. The interlocutors of the *Republic* candidly admit it. Yet they declare that fostering a belief along these lines will benefit the city (*Republic* 414b–415d). Those few who are wise and who know philosophy may grasp the true nature of the noble lie, but they will conceal it if they know what is good for their city. Which, of course, they do.

Many have questioned Plato's true goal in writing the *Republic*; it may not, after all, have been to offer a plan for an actual polity. Rather, the political order may be understood only as an allegorical portrait of an individual soul. Scholars taking the allegorical view often cite the utterly bizarre laws proposed for the ideal "state," laws very unlike those of any real-world polity at all. The reason these laws are so strange, the allegorists claim, is that they were not meant for application to a city, but only to a self. Numerological childbearing would obviously be a key exhibit in making this argument: An individual might marry for deep, inscrutable reasons of the soul, but he should not expect to command others to do likewise.

Indeed, by reading allegorically, even the noble lie becomes excusable: All of us have at one time or another resorted to tricks and expedients to overcome our lower, more bodily, and less-reasoned impulses. We find it noble to restrain such urges for the greater benefit of the self considered as a unity. On this view, the *Republic* is not to be understood as treating interpersonal politics at all. It is about the politics that takes place among the parts of the self. The interior of the self is a domain that is messy and complex, much like a polity, albeit with some radically different moral rules as well. The political allegory, then, brings into exceptionally stark relief the ethics of self-mastery. On this reading, the *Republic* is a description of how the self is *unlike* the state.

Yet an allegorical reading cannot so easily be given to the *Laws*. In the *Laws*, Plato offered what is probably his most definitive, and certainly his last, political prescription. And a prescription it most certainly is: Unlike the *Republic*, the *Laws* spends relatively little time discussing abstract theories of justice or individual psychology. The *Laws* is above all an instruction manual. It is straightforward, and meticulous, and one might even call it practical, were it not for the exceptional difficulty of doing what it commands. In the *Laws*, Plato described an interlocking system of written law and unwritten mores. The whole of it would govern essentially every aspect of individual life, and the rules were absurdly detailed.

There were, of course, numerous offices. There were to be "superintendents of the city streets, of buildings, private and public, of harbors, of the market, of springs, and not least, consecrated precincts, sanctuaries, and the like (*Laws*, 758e)." There would be "sacristans, priests, priestesses. ... city commissioners, and those who have control of the market (*Laws*, 759a–b)." The city guard would include "generals, taxiarchs, hipparchs, phylarchs, and prytanes" as well (*Laws* 760a–b); there were to be five "rural commissioners and captains of the watch (ibid)" from each of 12 tribes. Each of them would appoint 12 underlings. They were to oversee "dikes and trenches ... fortifications ... roads ... the flow of rain water ... dams ... exercising grounds and warm baths" among many other public works. The officers would also hear cases of dispute between citizens (*Laws* 760b–e).

The good government must also have "authorities in music and physical training," "superintendents of gymnasiums and schools" and "judges of performers contending in both musical and athletic competitions," and the latter will be "judges of both men and horses." There will be a "controller of the choirs" chosen by a two-tiered runoff election; the athletic judges will be appointed similarly (*Laws* 764c–765d).

Plato also fixed the number of households in his community. These, he wrote, "must remain forever unchanged, without increase or deviation whatsoever." The right number, for Plato, was 5040, again for apparently numerological reasons. Excess female children were to be married out of a household, and excess males were to be adopted out of one family and into another, so as to even out family sizes. When the population grew, the city should resort to sending out colonies, not merely to remain within reasonable limits, but to keep the population perfectly unchanging (*Laws* V 740b–e). Even in times of famine, plague, or war, when total households have been diminished below their appointed number, in no case were outsiders to be admitted (*Laws* V 741a).

Taken together, these were the most exceptionally stringent and inflexible laws on family life, immigration, marriage, and choice of domicile ever proposed. One also wonders how 5040 households could have supplied all the full-time officials that Plato's system demanded while still managing to feed itself.

There was more. The laws decreed the age at which men were to marry (*Laws* 721b; 772d–e). Unlike in the *Republic*, citizens of the *Laws'* community of Magnesia were to be allowed private, individual families, but they had to do so on a schedule. Marriage feasts were to be no larger than five friends and five kinsmen from each family (775a). There were regulations for the keeping of slaves (777c–e), for the plans of private houses, and for "the whole subject of [the city's] architecture in all its details" (778). There were to be rules as well for all of private and family life, although these were touched upon only incidentally:

The privacy of home life screens from the general observation many little incidents, too readily occasioned by a child's pain, pleasures, and passions, which are not in keeping with a legislator's recommendations, and tend to bring a medley of incongruities into the characters of our citizens. Now this is an evil for the public as a whole, for while the frequency and triviality of such faults make it both improper and undignified to penalize them by law, they are a real danger to such law as we do impose. (*Laws* 788a–b)

It was an important concession: Plato admitted that influencing mores, particularly among children, was much more difficult than legislating directly. Still, sometimes it was necessary all the same. The wise legislator would legislate for the heart as well as for everything else, and for exactly the same reasons.

And to what end? Plato answered this question with a reply worthy of the ancient historians:

In all that concerns city and fellow citizens, the best man, and the best by far, is he who would prize before an Olympian victory or any triumph in war or peace, the credit of victory in service to the laws of his home, as one who has all his life been their true servant above all men. (*Laws* V 729d–e)

The state exists to create excellent citizens; excellent citizens exist to preserve the state. They will do so, willingly, at the cost of their lives. The city aims to be eternal, because for Plato it is a sign of goodness to last eternally without any sort of change.

The dangers of decay were everywhere. In Book III, the Athenian—the speaker who served as Plato's mouthpiece—even claimed that the downfall of his own state began in the citizens' contempt for the *lawful forms of music*. From the abandonment of proper music, a whole parade of horribles followed:

[M]usic has given occasion to a general conceit of universal knowledge and contempt for law, and liberty has followed in their train. Fear was cast out by confidence in supposed knowledge, and the loss of it gave birth to impudence … So the next stage of the journey toward liberty will be the refusal to submit to the magistrates, and on this will follow emancipation from the authority and correction of parents and elders; then, obedience to the law, and, when that goal is all but reached, contempt for oaths, the plighted word, and all religion. (*Laws* III 701a–b)

One cannot be too careful, and so the *Laws* gives many long passages on good and bad kinds of music. These are to the modern reader almost impenetrable, because much knowledge about ancient Greek music is either obscure or lost. Ancient philosophers philosophized vigorously about music, but they did not play it, which was the role of slaves and women, and they did not leave descriptions sufficient for us to reconstruct it with much confidence. Within this now-lost art, however, Plato held that an exceptional danger lay in wait.

On this view, government may seem quite fragile. But while the threat of anarchy lurked in even the most apparently innocent activities, Plato also held that government was nonetheless everywhere and altogether *natural*. Plato even claimed (*Laws* III 680a–c) that the proverbially wild and uncivilized Cyclopes had "a form of polity." Paternal authority itself was the oldest form of law (θέμις); it was a sort of law even among the ostensibly lawless. Under paternal authority, other members of the family would "follow ... and form one flock, like so many birds, and [they] are thus under patriarchal control, the most justifiable of all types of royalty."

Eventually, the paternal authority of many different families would coalesce in the course of interactions among families, Plato claimed; those things that were agreeable to all would persist. Repeatedly, Plato invoked the metaphor of a shepherd and a flock of sheep, to stand for the statesman and the citizens of a polity (Book IV, 713c–3; V, 735a). The *polity* was to be the father of all, the shepherd of all, and it was to care for every human need. In this manner, an entirely artificial and almost unprecedented form of government acquired the appearance of naturalness.

Scholars have made various attempts to reconcile Plato's political writings: Some hold that the *Republic* described the polity intended for the most virtuous citizenry imaginable, while the *Laws* was meant for citizens who, while very virtuous, were less than ideal. I incline toward this view, although another group of scholars holds the reverse to be true; after all, the *Republic* contemplated somewhat laxer institutions for its lower classes, while the *Laws* held everyone to a stringent standard and made a number of significant extra prohibitions and regulations.

Whatever the case may be, for our purposes there is relatively little difference between the two communities. In both cases, government existed for similar ends, namely the preservation of virtue against the corruption of time, for which the individual was to sacrifice himself. Human states cannot be eternal or perfect, because nothing of this world ever is, but they should aim at the eternal and the perfect anyway, and they should deploy every possible resource in their quest.

Did Plato (or Socrates) think himself a statesman? And if so, did that mean actually legislating for this world, or was it only a matter of describing the ideal world, or describing heaven, or describing simply the well-ordered soul? Or did the statesman described in the dialogue perhaps not denote any corporeal person at all? Did Plato ever intend his prescriptions to be acted on? Or was the perfect city meant only to be contemplated, as Glaucon suggested near the end of the *Republic*, and as Socrates said was "probable"? (*Republic* 592b) Finally, as some have suggested, was the whole thing meant as an extended joke on politicians, much as Socrates had earlier nettled poets in *Ion* and the pious in *Euthyprho*?[15]

A more humane reading of Plato would not claim that the *Laws* was about real human beings, with our many differences, frailties, and often wonderful oddities. It might submit that Plato could not possibly have wanted to subject us all to such a demanding regimen. It remains conceivable that Plato's political writings were only an intellectual exercise, no more and no less.

Yet the vision of the beautiful, meticulously controlled, almost clockwork society has undeniably been Plato's legacy as a political philosopher. He has overwhelmingly been taken literally, whether he meant it or not, and the literal reading has had distinct consequences throughout both political philosophy and real-world politics. Alternate readings of the text cannot negate this fact.

The literal interpretation began early, too: Plato's most notable student, Aristotle, never hinted that Plato wrote allegorically or ironically in the political works. On the contrary, Aristotle's objections began by taking Plato at his word. For example, Aristotle condemned communal property and the community of women and children as tending to corrode sociability and friendship. He also observed, correctly, that the state described in the *Laws* would require an enormous territory to support its thousands of bureaucrats.[16] But nowhere did Aristotle suggest that Plato was *not doing politics*, or that his political works were allegories of the life of the mind, or that they were just philosophical jokes.

As we shall see, Plato's vision was destined to be taken seriously, and the best evidence we have is that this is precisely how he intended it.

As we have just mentioned, Aristotle (384–322 BCE) was a student of Plato's. Born in Stagirus, in northern Greece, he spent much of his career in Athens. He also served as tutor to Alexander the Great and accompanied him on his conquests in Asia. Aristotle's political works gave rise to a school or relatively practical and less utopian theorists. These would examine, as Aristotle himself had examined, the constitutional forms of

actually existing societies, and to theorize in reference to them. Yet while
Aristotle avoided the utopian temptations of his teacher Plato, his politics
remained strikingly collectivist.

For Aristotle, man was the *political* animal, one fit for associations with
others. One man all by himself was not properly speaking complete; nor
were a group of stateless men. And men only found their truest and fullest
expression within a certain type of political society, namely a *polis*. Aristotle
began his *Politics* as follows:

> Observation shows us, first, that every city [*polis*] is a species of association,
> and, secondly, that all associations come into being for the sake of some
> good—for all men do all their acts with a view to achieving something which
> is, in their view, a good. It is clear therefore that all associations aim at some
> good, and that the particular association which is the most sovereign of all,
> and includes all the rest, will pursue this aim most, and will thus be directed
> to the most sovereign of all goods. This most sovereign association is the
> city, as it is called, or the political association.[17]

For Aristotle, being part of a *polis* seems to have been necessary for an
individual to live the good life.

Aristotle's idea of the good was *teleological*, or ends-focused. For him,
nearly all good things were good *for the sake of* some other good thing: A
rake is good for gardening; a garden is good for producing food; food is
good for continuing one's life; and so on. Every good was part of a single
hierarchy, with lesser goods helping to further greater ones, all the way up
to the greatest possible good of all. For Aristotle, that greatest good was
a well-considered, well-founded human happiness that he termed *eudaimonia*. To count as good, our habits, material goods, and associations all
had to conduce to *eudaimonia* in one way or another. It was by this standard that they were all to be judged.

Aristotle further held that *eudaimonia* could *only* be had in the context
of a well-ordered *polis*. That made the *polis* a very high-order good indeed,
and Aristotle even asserted at one point that the state comes *prior to* the
individual and the family in the hierarchy of goods:

> [T]he city is prior in the order of nature to the family and the individual.
> The reason for this is that *the whole is necessarily prior to the part*. [emphasis
> added] If the whole [human] body is destroyed, there will not be a foot or
> a hand, except in that ambiguous sense in which one uses the same word to
> indicate a different thing. ... For if the individual is not self-sufficient when

he is isolated he will stand in the same relation to the whole as other parts do to their wholes. The man who is isolated, who is unable to share in the benefits of political association, or who has no need to share because he is already self-sufficient, is no part of the city, and must therefore be either a beast or a god.[18]

By "prior" here, Aristotle did not mean chronologically prior. He meant prior in the order of goodness, as the self is prior to the hand or the foot: We need the *polis* to be fully human.

So what is a *polis*?

That is actually a difficult question. Translators usually use the term "city" or "city-state" for *polis*, which is not a bad start. Ancient Greeks lived in cities that were relatively independent, small, and internally cohesive. Each was made up of not more than a few tens of thousands of male citizens, as well as a much larger population of individuals who did not enjoy the rewards of citizenship. The latter group included women, children, resident aliens, and slaves. Fifth-century BCE Athens is estimated to have had about 30,000 adult male citizens and a total population of about 250,000. It was the largest city-state at the time, both in citizen population and overall. In ancient Greece, the whole of the city was commonly called the *polis*.

To the philosophers, however, the *polis* often meant the *political association of citizens*. It could also mean *the whole body of the citizens*, and in both of these senses it specifically excluded all noncitizens.[19] To make matters more confusing, these three senses of the word *polis* appear to have coexisted, even within the text of Aristotle's *Politics* itself.[20]

Ultimately, though, Aristotle seems to have identified the *polis* most properly with the city's constitution, that is, with its political association. Both citizens and residents of other types are constantly being born and dying. The only constant element of the *polis*, the only thing that supplies its identity over time, was therefore the state and its constitution, which was prior to all else, and which belonged to the citizens.[21]

And yet when Aristotle wrote that *man* was a political animal, he wrote it knowing full well that most men did not participate in the *polis*. By his stated standards, nearly everyone would remain defective forever, much like a severed foot. Women and slaves for him did not deserve citizenship, because—for him—these groups were already naturally defective in reason and incapable of achieving the good life. And yet men, who might on his terms have had a sufficient endowment of reason, would often lack citizenship anyway. These men would fall short of perfection, not because of an inner defect, but because the *polis* excluded them.

This implication hit close to home. Aristotle himself was a *metic*, a resident alien at Athens; he lacked the citizenship that he so highly praised. Only in exceptional cases did the citizens confer citizenship on an outsider, and Aristotle never received it, his scholarly achievements notwithstanding.

The question of inclusion remains in our own day, and it bears significantly on the question of what governments are for. Aristotle's communal pursuit of the good has been an inspiration to communitarians across the centuries, but it should never be forgotten that it required, at least for him, a large population of individuals who could never enjoy its benefits. It was not merely that some would fall short of mankind's full potential; rather, for anyone at all to achieve the greatest good, others would have to be excluded.

Modern communitarians are properly shocked by this aspect of Aristotle's thought. Yet even our modern, egalitarian democracies struggle with a similar problem: In the end, there is only so much rulership to go around. If the good consists of active participation in the government to an extent comparable to an Athenian citizen, then almost none of us are fully good. Almost none of us can be, because only a few can enjoy expansive civic authority.

Adding more rulership does not seem to be a workable solution. Even if we subdivided the United States into *polis*-sized communities, it would still not supply enough: The *polis* itself excluded so many, and still it supplied only so much rulership. Worse, at least to our modern sensibilities, those who actively govern us do not appear more moral or virtuous than the rest of us, and multiplying their number seems likely to be unwise for other reasons.

But back to the past: Did the good life of the *polis* refer to the life of the citizens as a whole, including the parts we think of today as their private life? Or did the *polis* only mean the part we nowadays call the state? (Did the Greeks have a clear idea of society that was wholly abstracted from government, or not?) It is instructive to consider that the English word "city" has a similar ambiguity: We say that a city has a population of ten million, but then, the city also collects trash on Tuesdays.

Did the ancients have a conception of social life that was removed from, and independent of, the polis? Philosopher Roderick T. Long writes that for Plato, and also to a significant degree for Aristotle, "society and state were merged into one entity, the *polis*—a term which ... cannot be translated as either 'society' or 'state,' since it was both. The *polis* ... was an organic community whose authority governed every aspect of life."[22] For the ancient philosophers, the *polis* was everything, then, and private life was for them very little, or possibly even nothing at all (certainly it was nothing to Plato).

Long insists, however, that not all Greeks thought this way. He holds that Plato and Aristotle were unusual and described "ideals that they saw as antithetical to Greek society as it actually existed." He finds evidence of a flourishing private life in ancient Greece of a type that we might call *civil society*. Long finds ancient Greek civil society to have included private banking, private charity, private religious observances, and of course private friendship, one of the few institutions that the state has never effectively displaced. Long argues that even back then, civil society was in many ways preferable to politics. Modern Aristotelians, Long among them, have gone on to question whether the state *deserves* the lofty place that Aristotle gave it, even as they have defended many other Aristotelian ideas.[23]

I am personally not as convinced that Plato and Aristotle were so strange for their time and place. The evidence of the ancient historians, given above, suggests that their views were perhaps altogether conventional, the functional importance of civil society notwithstanding. It is also possible that this expansive view of the state was common only among a highly educated caste of scholars, including both philosophers and historians. Given that these were the most literate members of society, their works are what have survived down to the present. creating an inherent selection effect that is difficult to avoid. If so, it would hardly be the last time in history that the academy held views antithetical to those that sustained the society at large.

Ultimately, though, much evidence suggests that the expansive view of government was current among the well-educated. We moderns may read Aristotle in a more liberal vein, and we might even try the same with Plato, yet I suspect that few of Aristotle's or Plato's contemporaries would have shared our interpretations. Both may remain interesting and important thinkers without having to become moderns in all respects.

It is a further question whether Aristotle would have seen anything like his *polis* in our modern political communities. Our nation-states, and especially our transnational communities, are vastly larger and more diverse than any of the city-states he took as the natural units of politics. It is likely he would have mistrusted our polities simply for their size. The Greek *polis* was often hardly more than a small town by today's standards, and in

Aristotle's time the political communities closest in size to ours were the empires of the east, which Aristotle and almost all other Greeks despised.

We now have a very specific and very qualified claim: Aristotle believed that mankind was fit for political association, but that a small share of men will actually enjoy it. These men will become excellent through politics. This specifically political excellence does not come from any of the other also excellent things that may be found in or around the *polis*, such as literacy, religion, commerce, the family, or friendship.

Given his own status, was Aristotle trying to flatter the citizens around him? Are city dwellers really more excellent in the first place? In identifying the good with civic participation, might one argue that Aristotle seized on a coincidence? Yes, the residents of a *polis* may be said to have been good, but might that have come from friendship, education, or culture, rather than from political participation?

Now, these possibilities are all difficult to disentangle from the presence of politics. Historically, many other advantages have almost always coincided with the presence of political institutions, which are always strongly in evidence in densely populated areas. Causal inferences are dangerous, then, and we should not rely too strongly on them in a theory purporting to offer the purpose of government or to fix our relationship to it.

Might Aristotle have suffered from a *pro-political bias*? It seems likely: As we shall see throughout this study, it is reasonable to suspect that many other authors, from the privileged to the marginalized, all converged on a too favorable view of politics, whether because they sought rulers' patronage, or because they feared rulers' anger, or, in Aristotle's case, a measure of both.

Supposing Aristotle was right, however, and the political association really is the highest social good, and we really do stand in relation to the *polis* as parts to a whole. If so, then we should expect to see that our own excellence is only found in political communities. On investigation, that might turn out to be the case—*polis* life might turn out to produce better human beings. But to observe as much and to offer it as evidence for the claim that we are in fact mere parts of a greater whole is a mistake that Aristotle should never have made. After all, among his many other contributions to philosophy, he was the first to have described the fallacy of affirming the consequent.

Note also that this view of the good—which places the state as the highest and most important social instrument for achieving the individual good—is in some conflict with the philosophical, contemplative life

Aristotle recommended in the *Nichomachean Ethics*. There he held that the life of the statesman is the *second best* life to be had, after that of the philosopher: Yet if the state is the highest of all human associations, and if we are not even properly human without it, it is hard to understand how the statesman does not aim at the good more truly than even the philosopher.[24]

Indeed, Aristotle's approach to ethics more generally is remarkably individualist in its orientation. It counsels the cultivation of the self and of the individual virtues, and it always looks to actuate potentials within individual selves through a process of largely individual effort. Aristotle's ethics—to say nothing of his epistemology or logic—treat of individualized processes. It is remarkable how little his political philosophy follows suit.

A more individualist approach to politics that still builds on Aristotelian ethical foundations might hold that the state is *instrumental to flourishing*, in a manner analogous to medicine. An unhealthy person necessarily falls short of human excellence in one key aspect, but it would be nonsensical to claim that his defect stemmed from failing to join the whole (medicine!?) of which he was merely a part. On the contrary, one might say he should take his medicine and then be done with it as soon as he properly can.

In like manner, the state may solve specific problems for us. The state therefore does contribute to human excellence, albeit in an instrumental way. A good physician will heal, of course, but he will also have some regard for our autonomy, and he will take care *not* to treat us in ways that render us merely a part of a larger and probably undesired whole. We would resent and find threatening a doctor who wished to attach us permanently, Borg-like, to a collective. The state could very well be like *that* doctor, normatively speaking. But if so, we should consider it an evil that much of western philosophy has not yet apprehended.

NOTES

1. Plutarch, *Moralia*, "On the Glory of the Athenians," Loeb Classical Library, Vol. IV, 1936, 347d.
2. Lucian of Samosata, "A Slip of the Tongue in Salutation."
3. Even despite some further changes: Our modern marathon is 26 miles, 385 yards, which is the distance from the 1908 London Olympic Stadium to Windsor Castle.
4. Herodotus. *Histories*, 1.30.

5. Polybius. *Histories*, 1.1.
6. Livy. *History of Rome*, 1.1.
7. Plutarch. *Lives*, Ch. 6, 21.
8. Ibid., Ch. 45.
9. Cicero in *De Re Publica* writes that Archimedes built two devices that may have been similar to the Antikythera mechanism, but the description that Cicero gave was cursory. The devices in question seem to have been mechanical planetariums or orreries, rather than portable indicators. Cicero, *Political Works*. http://oll.libertyfund.org/title/546/.
10. Morley, "Trajan's Engines."
11. Varro, *De Re Rustica* I.12.2: "Precautions must also be taken in the neighbourhood of swamps, both for the reasons given, and because there are bred certain minute creatures which cannot be seen by the eyes, which float in the air and enter the body through the mouth and nose and there cause serious diseases." Varro's errors in biology, and they are many, admittedly compromise the otherwise striking effect of this passage.
12. *Statesman* 292e; 294a, in *Plato: The Collected Dialogues*. Unless otherwise noted, all citations to Plato are to this volume.
13. Plato, *Republic*, 546b–c. The "pempad" is simply the number five, dressed up in obscurantist language, the English of which is faithful to Plato's obscurantist Greek. See Carl A. Huffman, "Mathematics in Plato's *Republic*," esp pp. 219–220.
14. Karl Popper, *The Open Society and Its Enemies* vol 1. New York: Routledge Classics, 2003, p. 85.
15. See Bloom, *The Republic of Plato*.
16. See Aristotle, *Politics*, II.1–6.
17. Aristotle, *Politics*, 1252a.
18. Ibid., 1253a, p. 11.
19. Hansen, "Was Athens a Democracy?"
20. In *Politics* Book III, chapter 1, Aristotle discusses how the city, the state, and the aggregate of citizens are sometimes disunited as to their actions and purposes. See 1274b32–1275a19.
21. Aristotle, *Politics*, III.3
22. Long, "Civil Society in Ancient Greece."
23. Rasmussen and Den Uyl, *Norms of Liberty*, is a good place to begin on the topic.
24. Aristotle, *Nichomachean Ethics*, X.7.

CHAPTER 3

The Ancient Dissenters

Even in ancient times, however, some dissented from the idea that man is fitted for the *polis*, and that he does best by obeisance to it. These dissenters often formed something of a fringe group, to be sure. But they did exist.

Among the first of the dissenters was Diogenes the Cynic (c.412–323 BCE), who famously lived in an empty barrel and scorned all material wealth. Once, when Diogenes was sunning himself, it is said that Alexander the Great "was standing by, and said to him, 'Ask any favour you choose of me.' And he replied, 'Cease to shade me from the sun.'"[1]

Clearly the anecdote reflects Diogenes' contempt for material wealth. Many would surely have demanded cash, land, or power, which Diogenes did not. A second reading, however, has political implications: Diogenes may have been suggesting that rulers, and perhaps government itself, are impediments.

Alexander attempted to exit gracefully, saying, "But verily, if I were not Alexander, I would be Diogenes." To which some sources add that Diogenes answered: "If I were not Diogenes, I should also wish to be Diogenes."[2] And by implication: not Alexander.

Other incidents from the life of Diogenes underscore his skepticism about government. Once, when asked what city he came from, Diogenes replied: *I am a citizen of the world*. Or: *I am a cosmopolitan*.

In our modern idiom, these words have a clear meaning. They mean that the speaker has no strong allegiance to any nation. A cosmopolitan has probably traveled a lot, read widely in several languages, and may have

© The Author(s) 2017
J. Kuznicki, *Technology and the End of Authority*,
DOI 10.1007/978-3-319-48692-5_3

many foreign friends and contacts. A cosmopolitan could live in Paris or Tokyo, and things would go just as well either way.

This sort of cosmopolitanism is nothing new to us. It's also not what Diogenes meant. By calling himself a cosmopolitan, he declared himself to have worth and dignity not by virtue of local, particular laws—but by virtue of a higher law, one that was universal to all. Being universal, this law perhaps even included such outside groups as slaves, barbarians, and women.

Another anecdote of Diogenes' life demonstrates the point. Diogenes was once captured and sold as a slave; when asked by the slave trader what work he could do, "he bade the crier 'give notice that if any one wants to purchase a master, there is one here for him.'" Yet if a slave could be a good master, then the difference between slave and master was one of mere happenstance. Such a difference couldn't possibly be just. Rather, all people deserve to be treated as equals in worth and dignity.[3]

In the ancient world, all of this must have sounded strange and maybe even comical. Recall how narrow the class of citizens was even in "democratic" Athens. To claim citizenship in all places at once was not simply to declare one's worldliness or interchangeability. It was to stake out a radically different moral foundation from one's peers. It was to stand apart from, and to challenge, nearly everything about Greek society.

Diogenes wasn't entirely alone, however. Another early cosmopolitan was Antiphon the Sophist. Antiphon was a contemporary of Socrates who opposed many aspects of Socratic philosophy, above all its politics. Little is known about Antiphon's life; indeed, Antiphon the Sophist may even be identical to Antiphon of Rhamnus, a talented contemporary orator (Modern scholars favor uniting them, but they aren't completely certain). In any case, only a few fragments of Antiphon's philosophical works survive. But they offer an intriguing vision of a higher moral law, one that was held to be binding on all people. As one of Antiphon's relatively intact fragments put it,

> We [respect] and revere those who are of good parentage, but those who are not of good family we neither [respect] nor revere. In this behavior we have become like barbarians one to another, when in fact by nature we all have the same nature in all particulars, barbarians and Greeks. We have only to consider the things which are natural and necessary to all mankind. These are open to all [to get] in the same way, and in [all] these there is no distinction of barbarian or Greek. For we all breathe out into the air by the mouth and the nose, and we [all eat with our hands]...[4]

If anything, Antiphon made a stronger claim than Diogenes: After all, Diogenes only claimed *personally* to be a citizen of the world. Conceivably, this might have stemmed from some unique quality of his—his genius, perhaps. With so little to go on, Diogenes might conceivably *not* have been arguing for a universal idea of justice after all—he might have been saying only that he was an exceptional man. That's not such an interesting claim. Or even an original one.

But Antiphon called out his fellow Greeks. He claimed they made barbarians out of themselves, a serious insult. Not only that, but he suggested that *two* social distinctions should fall: one between slave and citizen, and the other between Greek and barbarian. The true barbarians, he suggested, were not the ones who were commonly *called* barbarians. The true barbarians were the ones doing the name-calling.

Antiphon's argument was fundamentally about justice. One of its implications was that government, if it is to justify itself, must be justifiable to all, and not merely to Greeks. And government, if it is to have *any* purpose, must have a purpose that treats Greeks and barbarians alike. It was a radical critique of both citizenship and government as they then existed.

Admittedly Antiphon held other, less sympathetic, views. He appears to have held that justice consisted above all of never harming anyone who had not personally harmed you. He therefore condemned the acts of testifying and presiding at a trial: Each might result in the defendant getting hurt by a state functionary. This would break his first rule of justice. In Antiphon's view, perhaps no state could ever be legitimate, because government may always require harm by third parties. Antiphon may well have been an anarchist, one who rejected all forms of agent-administered justice—a step even most modern anarchists refuse.

Annoyingly, though, it's hard to be sure. Antiphon's extant writings are often crabbed and frankly preposterous. They are also so incomplete that we cannot say for certain whether these were his final views, or whether he was expounding them so as to refute them later on.[5] There is good reason why Socrates *is* Socrates, and why Antiphon is just another sophist. But sometimes Antiphon appears to have scored a direct hit. When he objected to the narrow, parochial, and yet all-demanding city-states of ancient Greek philosophy, Antiphon anticipated a good deal of later opposition to the claims of the all-powerful state.

Still, Antiphon seems to have had fairly little influence on later philosophers. Diogenes did a bit better on that score: The influential school known as Stoicism traced its origins to him. Like him, the Stoics disdained material wealth and temporal power, holding that the moral good was the only truly good thing in the world. A wise man would moderate his emotions, said the Stoics, above all when they sprang from transient gains or losses. The wise man might do best to retire from or even disdain public life. Rather than seeking personal glory in politics or on the battlefield, he could refer to eternal truths for his consolation.

Stoicism was a popular philosophy in the Roman world, and notable Roman-era Stoics ranged from Epictetus, a Greek who was born a slave, to the Emperor Marcus Aurelius, who (perhaps surprisingly) embraced the Stoic philosophy of one of his tutors, Quintus Junius Rusticus. The Stoic most familiar to the modern world is probably Marcus Tullius Cicero (106–43 BCE), one of ancient Rome's greatest orators. Cicero clearly articulated the moral ideal of a shared, universal humanity, one that might encompass a slave and an emperor alike. Such an ideal would—logically, at least—limit the scope of state powers, because it would bind everyone equally. It would also limit the particular duties of individuals who were subject to various governments.

Cicero's work *De Finibus* was a key defense of Stoicism. In it, he wrote,

> [W]e derive from nature herself the impulse to love those to whom we have given birth. From this impulse is developed the sense of mutual attraction which unites human beings as such; this also is bestowed by nature. The mere fact of their common humanity requires that one man should feel another man to be akin to him.[6]

Cicero then asserted that our *common humanity* was the real reason for government; states were not properly formed, or properly defended, simply to win a local or particular advantage. Such actions could not be justified with reference to our common humanity. Rather, they appear to imply particularity.

For Cicero, the likeness of one human to another, and the common origin of all of us, was enough to give humanity a common natural law. To break that natural law was to act wrongly, no matter what one's social status, and no matter what a government may have ordered. At their best, written laws captured the natural law in words and informed citizens of their duties; at their worst, written laws contravened natural law and therefore did harm.

It was a bold stance, but one that Cicero seems to have substantiated throughout his other works. He did not hesitate to make powerful enemies in Rome, including Julius Caesar, whose death he heartily approved of. He naturally disdained Mark Antony, who led the Caesarian political faction, and whom Cicero believed to be subverting what little remained of the republic. Antony had Cicero assassinated.

Finally, many also see a cosmopolitan strain in the *Elements of Ethics* by the later Stoic philosopher Hierocles. We know almost nothing of Hierocles' life, except that he lived in the second century CE. Fortunately, some of his writing survives. Hierocles proposed that we all lived within concentric circles of affection:

> For the first, indeed, and most proximate circle is that which every one describes about his own mind as a center, in which circle the body, and whatever is assumed for the sake of the body, are comprehended. For this is nearly the smallest circle, and almost touches the center itself. The second ... [which] is at a greater distance from the center ... is that in which parents, brothers, wife, and children are arranged. The third circle ... is that which contains uncles and aunts, grandfathers and grandmothers, and the children of brothers and sisters. After this is the circle which comprehends the remaining relatives. Next to this is that which contains the common people, then that which comprehends those of the same tribe, afterwards that which contains the citizens; and then two other circles follow, one being the circle of those that dwell in the vicinity of the city, and the other, of those of the same province. But the outermost and greatest circle, and which comprehends all the other circles, is that of the whole human race.[7]

He recommended habits that he thought would draw these circles closer and closer together. In this way, we would fulfill our duties toward the ones we love, as well as making it easier for others to do likewise. Because each of us lives within a similar series of concentric circles, the habits of a good life would draw humanity together.

Taken together, the Stoic tradition offers an important corrective to the otherwise all too common tendencies of hierarchy and of particularism, both of which have been taken for legitimate purposes of government: Something else, says Stoicism, something other than the pursuit of narrow, in-group interest, is what humanity should aim at. That thing is the life of virtue, which all may equally pursue, regardless of personal circumstances. As philosopher Mark LeBar puts it, "this 'cosmopolitan' dimension of Stoic thought seems to dissipate entirely the idea that there is something important in political justice, as the bounds of the polis seem not to be relevant for thinking about morality generally or justice in particular."[8]

But Stoicism offered only an interior, intellectual freedom. Especially in its later period, Stoicism saw little prospect for realizing an enduring political or social change. Its adherents schooled the ancient Roman elite, but Stoicism did not stir them to question the institution that they directed. Government was to be suffered with patience; whatever universal equality existed among men in theory remained absent in practice, and its absence was simply to be endured. Unlike Aristotle, who held that properly used material goods could indeed conduce to the good life, the Stoics disdained the material world entirely. As a result, Stoicism had virtually no sensible economics, for example, and no foundations on which to bother building one. Its social commentary was therefore to some degree inherently limited.

Finally, let us consider one other group of dissenters. These were the followers of Epicurus. In many ways, their philosophy will be familiar to present-day readers. The Epicureans were materialists; they denied the reality of the soul, the gods, and the afterlife. Epicureans also counseled that freedom from worry was the purpose of philosophical inquiry. This freedom could be had by understanding that essentially all commonly experienced fears were about limited, ephemeral, or even nonexistent things: Pain can be endured, and it does not last forever. Material wants can be satisfied, often more easily than we imagine. The gods are not to be feared, because they do not exist. And death is not to be feared, because death is not an eternal suffering. It's only a brief suffering, followed by nothing at all, and it would be silly to experience fear over nothing.

For the Epicureans, justice was an agreement not to harm others, for the sake of not being harmed by them in turn. Justice, then, served the Epicureans' goal of being free from worry; justice meant removing the specific worry that others would harm you. As such, it was neither arbitrary nor illusory; although it was a human creation, justice was to be desired as a part of the good life. As Epicurus wrote,

> There never was an absolute justice, but only an agreement made in reciprocal intercourse in whatever localities now and again from time to time, providing against the infliction or suffering of harm.
> Taken generally, justice is the same for all, to wit, something found expedient in mutual intercourse; but in its application to particular cases of locality or conditions of whatever kind, it varies under different circumstances.

Those who were best able to provide themselves with the means of security against their neighbours, being thus in possession of the surest guarantee, passed the most agreeable life in each other's society.[9]

Epicurus was trying to solve a difficult problem: At times, obedience to a mere convention will work out to everyone's advantage. And yet justice isn't simply whatever we make up or imagine; we can easily think of unjust conventions, but we shouldn't adhere to them. So how do we choose among conventions? Worse, a philosopher can readily appreciate that our current conventions are neither eternal nor transcendent. How, then, can they be part of the good, which is commonly affirmed to be both?

Epicurus proposed that the solution to these problems lay in the *universal* desire to be free from worry, coupled with the *particular* agreements that are deployed to that end. For example, in our own time, the practice of driving on the right-hand side of the road is not to be derived philosophically. It's not eternal or ordained by God. It's not even universal in practice, and it clearly rests on a mere human agreement. Yet having to renegotiate the proper side of the road at every meeting of two cars would be costly, dangerous, and time-consuming. Agreeing on one of two possible conventions, left or right, will work out to the greatest advantage for everyone involved. Rational argument would appear unable to decide between them. And yet when we have made our choice, we can say that it is a part of justice, because it conduces to the freedom from worry.

Taken together, the philosophical dissenters of the ancient world undoubtedly consoled many reflective people. But they offered little prospect of, or inspiration for, social change. In particular, Stoicism often called for withdrawal or detachment from politics, while Epicureanism was frequently caricatured and dismissed (notably by Cicero), despite offering what may have been the most promising theory of government anywhere in the ancient world.

In the centuries to come, Aristotelians and Platonists predominated intellectually, along with others who extolled self-sacrifice for the sake of the state. Platonic and Aristotelian philosophy both placed the state at the center of society. The ancient historians readily concurred, urging all men to a life of typically bloody self-sacrifice. The consensus view of the good state was not that it was an agreement, but that it consisted of mastery over others, with or without their consent. Skepticism about the use of government, and attempts to survey and limit its power, got a rough start in western philosophy.

NOTES

1. Diogenes Laertius, *The Lives and Opinions of Eminent Philosophers*, "Diogenes."
2. Plutarch, *Lives*, ch 14.
3. Diogenes Laertius, *The Lives and Opinions of Eminent Philosophers*, "Diogenes."
4. Antiphon the Sophist, *On Truth*, Fragment B 91. In *The Older Sophists*, pp. 220–221.
5. Perhaps the best word to describe Antiphon, or at least what we still have of him, is *antinomian*. *Nomos* typically meant the man-made, written or statutory law. Antiphon mostly or completely rejected *nomos*. He held that the crafting of *nomos* was error-prone, the administering of *nomos* was unethical—for the reasons described above—and the rewards and penalties of *nomos* were inadequate and uncertain in their working. *Nomos* was therefore unjust. See Ostwald, "Nomos and Phusis."
6. Cicero, *De Finibus*. ss. 62–63.
7. *Political fragments of Archytas*, pp. 106–07.
8. LeBar, "The Virtue of Justice, Revisited."
9. Diogenes Laertius, *Lives of Eminent Philosophers*, "Epicurus."

CHAPTER 4

Christianity and the City of Man: From Retreat to Reform

Saint Augustine (354–430 CE) was the Catholic Bishop of Hippo, a city in what is now Algeria. He is best known as a theologian, but here we will consider him as historian and a political philosopher, which he was as well. His answer to the question "What is government for?" was startling and original, and it remains vital right down to the present day.

In 410, the Visigoths sacked Rome. And in the previous centuries, a new religion, Christianity, had taken hold across the empire and beyond. In his enormously influential work *The City of God*, Augustine sought to explain what both of these events really meant—and how they related to one another. His effort arrived at a turning point for the history of western political philosophy. After more than 800 unconquered years, Rome turned out to be *temporary*. The imperial city, the ruler of the world, had been overthrown. Did it really make sense to sacrifice oneself for this kind of worldly glory? (Not anymore? Or did it never really make sense at all?) If the state wasn't a theater of virtue, then what was it doing for us?

Not only was Rome temporary, but so were the old gods, whose cults had supported the state. Those gods had either failed or abandoned Rome, and no one could sensibly deny it. Understandably enough, Augustine's response was that worshipping the old Roman gods had been foolish and ineffective. They could not supply earthly safety, and still less could they give hope for eternal life. But the one God who made all things on earth *had* promised us eternal life. Incarnated in the person of Jesus Christ, he called all of us to His worship. In contrast to the temporary City of Man, Augustine described the City of God as follows:

© The Author(s) 2017
J. Kuznicki, *Technology and the End of Authority*,
DOI 10.1007/978-3-319-48692-5_4

a city surpassingly glorious, whether we view it as it still lives by faith in this fleeting course of time ... or as it shall dwell in the fixed stability of its eternal seat, which it now with patience waits for, expecting until righteousness shall return unto judgment, and it obtain, by virtue of its excellence, final victory and perfect peace.[1]

Let the cities of the world fall, Augustine declared; a Christian would have his reward in Heaven. Even the recent Christian emperors could only be said to have had happiness on earth by virtue of their hope in the kingdom to come. And:

[T]rue justice has no existence save in that republic whose founder and ruler is Christ, if at least any choose to call this a republic ... in this city is true justice, the city of which Holy Scripture says, Glorious things are said of you, O city of God.[2]

Augustine's condemnation of pagan Rome was fierce:

With respect, then, to moral evils, evils of life and conduct – evils which are so mighty, that, according to the wisest pagans, by them states are ruined while their cities stand uninjured – their gods made not the smallest provision for preserving their worshippers from these evils, but, on the contrary, took special pains to increase them.[3]

Where the pagan Romans had prided themselves on their civic virtue, Augustine denied it entirely. Augustine was a devoted student of Roman history, and he walked his readers through many examples of depraved Roman conduct, from the earliest days of the city to his own time. He culminated his narrative with accounts of some of the lewd and shocking rites performed in certain Roman temples—rites he had personally observed, much to his own later shame.

Indeed, *The City of God* is one reason why imperial Rome has always enjoyed a sulfurous reputation. Augustine spared no corner of the Empire's conduct. He even likened imperial valor in battle to imperial gluttony:

Do they reply that the Roman empire could never have been so widely extended, nor so glorious, save by constant and unintermitting wars? [...] In this little world of man's body, is it not better to have a moderate stature, and health with it, than to attain the huge dimensions of a giant by unnatural torments, and when you attain it to find no rest, but to be pained the more in proportion to the size of your members?[4]

Rome was not grand. It was bloated. Worse, given the injustice of the Roman system, the Roman state itself amounted to little more than thievery:

> Justice being taken away, then, what are kingdoms but great robberies? For what are robberies themselves, but little kingdoms? The band itself is made up of men; it is ruled by the authority of a prince, it is knit together by the pact of the confederacy; the booty is divided by the law agreed on. If, by the admittance of abandoned men, this evil increases to such a degree that it holds places, fixes abodes, takes possession of cities, and subdues peoples, it assumes the more plainly the name of a kingdom, because the reality is now manifestly conferred on it, not by the removal of covetousness, but by the addition of impunity.[5]

In this passage, Augustine followed Cicero, who had made similar observations[6]; he also anticipated some aspects of the field of economics that has become known as *public choice theory*. Nobel laureate James Buchanan once described public choice theory as "politics without romance."[7] It attempts to explain state actions by treating state agents as self-interested economic actors, and not as agents morally distinct from the rest of us. To public choice theorists, state actors are not idealized statesmen, not the agents of God, and not the harmonious members of something like a human body. Public choice seeks, rather, to explain state agents' actions by self-interest alone; most commonly, political acts can be understood as efforts to preserve politicians' own power and respect.

Thus in his book *Power and Prosperity: Outgrowing Communist and Capitalist Dictatorships*, the eminent public choice economist Mancur Olson likened states—of all kinds—to "stationary bandits": Although taxation may be confiscatory and wholly unjustified, one might still have (at least a slight) preference for the state, the stationary bandits, over the roving kind: The state generally confiscates only at regular intervals, and it does at least take care to keep other bandits out.[8] It is not a flattering picture of state, but it is a lot like what Augustine had written. Without justice, where exactly is the difference between a prince and a robber?

Yet how did it happen that such wicked people held dominion over the entire known world? God, claimed Augustine, had permitted Rome to stand, but only out of punishment for human evil, and not because of any genuine virtues of the Romans. "God," wrote Augustine, "gives kingly power on earth both to the pious and the impious."[9] To explain the Empire's evil, Augustine offered a two-pronged argument: Either the Roman gods were demons and devils—in which case they had drawn mankind into sin and suffering—or else the Roman gods were much-embellished historical accounts of actual people—in which case, Rome had only itself to blame. Either way, Augustine called on his readers to coexist with Rome while understanding themselves to stand spiritually outside of it. Unfortunately, there is no physical escape from Rome while living in the earthly realm: As modern-day political philosopher Jacob T. Levy puts it, as long as we live on Earth, "God lets us live in the mess that we made."[10]

So what are the few good people supposed do while down here on earth? Says Augustine,

> [T]he people of Christ, whatever be their condition ... are enjoined to endure this earthy republic, wicked and dissolute as it is, so that they may by this endurance win for themselves an eminent place in that most holy and august assembly of angels and republic of heaven, in which the will of God is the law.[11]

When compared to Aristotle, one can't imagine a more different relationship to the earthly state. One does *not* participate in the earthly state to realize one's true nature or to act out virtue. On the contrary, a virtuous person *endures* the earthly state. A virtuous person *suffers* it, with patience, as one might suffer a disease.

The resemblance to the Stoic philosophy is strong, but not total. For Augustine, participating in a worldly state would be morally reprehensible; this type of state at the very least is cruel, rapacious, perverted, and pagan. Unlike the Stoics, however, Augustine promises that eternal happiness awaits, with thanks to God's grace, provided we keep our hands clean and remain humble and pious in all things. There is no question whatsoever of controlling the state, at least for the time being. The Empire is, ultimately, a cross to bear. Or, as Levy puts it, "Politics *happens to us.* We don't will it into being."[12]

For the vast majority of the Empire's subjects, this was indisputably the case. It had been true for adult male Roman citizens, and even for Roman Senators, at least since the time of Julius Caesar, some 450 years earlier. But above all, politics had happened to the Christian minority, who had experienced sporadic but decidedly cruel persecutions ever since 64 CE. In that year, Nero blamed the Christians for the Great Fire of Rome, which he may personally have started. He then used the Christians' supposed crime as a pretext for mass executions.

By the fourth century, however, the fortunes of Christianity had risen considerably. Christianity was officially tolerated in the Empire beginning under Constantine in 313, and it became the state religion, to the exclusion of all others, under Theodosius in 393. Still, its situation was evidently at least a little precarious, particularly in the West: Christianity was in the ascendancy among the upper classes, but traditional Roman paganism remained common among the lower classes and in outlying areas. It could conceivably have come back. The early books of *The City of God* are even devoted to rebutting the charge that Christianity was responsible for the Visigoths' sack of Rome, a charge that could obviously cause a new faith to waver. Politics might still have happened to the Christians again, even in Augustine's time.

As subjects of a collapsing state, Augustine's readers perhaps found solace in their hope for the life to come, and in a healthy disengagement from the state, which they could no longer consider a locus of virtue. To Christians at least, Augustine offered a particularly satisfying way of thinking about the follies and evils of government, which everyone had to suffer. It's also easy to see how his stance might have helped Christianity win converts and eventually prevail: It offered a way for believers to separate themselves from the dramatic reversals of fortune that they saw all around them.

We must resist, however, the temptation to conclude that Augustine was a libertarian in the modern sense. He was no such thing. Were it in his power, Augustine would have instituted a Christian theocracy. He would not have attempted to craft a liberal state, and he would certainly not have abolished the state altogether. Augustine clearly believed that divinely ordained government was possible on earth. Yet intellectually it is only a small step from Augustine's arguments about Rome to the conclusion that government itself is always an affliction. All that is lacking is a skepticism about the prospects of practical theocracy.

We might also question the perhaps too-easy conclusions Augustine draws about the Empire being merely a cross to bear with patience. His view may comfort Christians, but it will not comfort those of other religions or of no religion, who are not accustomed to bearing crosses, and who may find no moral worth in doing so. And even for Christians, it is unclear exactly why government should be a cross to bear rather than an affliction to assuage. After all, Jesus did not counsel the hungry on the virtues of fasting. He fed them. In our own day, other burdens that were once thought divinely ordained have likewise been lifted. Plagues can be cured, childbirth can be relatively painless and, at times, justice can even be done. Why not do justice more often? Why not restrain the state?

Augustine nonetheless raised a serious question, even if he did so inadvertently: What if something like the Empire is all that politics has to offer? What if there was no way to achieve the City of God on earth? If everyone took this view to heart, and if everyone *always* shunned states as impious, the result could be a pacifist Christian anarchy, akin to the one described by Leo Tolstoy.[13] Augustine did not expect or hope for this outcome. He took it for granted that reprobates would always exist. And he knew from Roman history that reprobates would always be drawn to power. The way to stop them, he believed, was to set up a Christian government.

What about today? Are modern western democracies more like Aristotle's idea of the state, or are they more like Augustine's? Is government today an instrument of our will? Or is government more like a cross to bear?

It's tempting to say that the truth lies somewhere in between: At times, we experience the state as something to be endured. Thankfully, in a liberal democracy, our endurance almost never takes the form of literal martyrdom, but the state still certainly frustrates our individual wills concerning a wide array of choices that we might otherwise have made. This happens to everyone, not simply to anarchists, and it cannot be easily denied (True, some philosophers have done so, including Rousseau and Hegel, whom we will discuss later).

Still, there are some times when we feel that the state has done precisely what we would have done, if we had had the power to direct its actions. In these times, we may want to claim that the state has been an instrument of our will. From such times, we may even want to generalize, and to say that in a liberal democracy, government *does* carry out our will. We might even come to think, with Rousseau, that it had enacted a sort of General Will.

This would be a mistake. At best, the state's actions may *coincide* with our will. But the state is not in fact *responding* to our will. It's responding to a large aggregate of wills, and, for the state's purposes, it makes no difference whatsoever whether our particular will happens to be part of the aggregate. The state would do just the same in either case.

Claiming credit for the state's actions is a lot like claiming credit when a favorite sports team wins. Well-known processes brought about the victory; whatever an individual fan might have willed has had no effect at all. So too with the actions carried out by a democratic government: I may happen to *like* what the democracy does. But I would have ended up equally governed either way.

In a democracy, state actions are almost never responsive to individual wills. And this is not by accident, but by design. Democracy could almost be defined as a type of government that responds *as little as possible* to any individual will. Democracy attempts rather to realize a collective will through the electoral process; it systematically acts on this collective will over the will of any individual. Broken into suitably small chunks, state actions in a liberal democracy can and of course do command voting majorities, and it is from this fact that the collective will of the people is inferred, perhaps for lack of any better approach. But when the state acts in this way, the individual wills of those in the minority are always violated, and all of us are minorities on some point or another. Meanwhile, those in the majority experience no more than a happy coincidence, because nothing that the state did had ever depended on their particular will: Willing otherwise, even in conjunction with many others, would not have made it so. In that sense, politics still happens to us all.

And yet enduring the actions of the state may be preferable to the next best alternative. That alternative may entail submitting to the mere arbitrary will of one other person, or even experiencing a constant state of insecurity. If that's the case, then having one's will frustrated on some other matters may constitute an improvement. It's not clear, however, just how often we *ought to* exercise our capacity for endurance of the state, rather than finding some other, and more voluntary, methods of attaining a desired end. This is a central problem, I think, in trying to understand Augustine's politics. It's also a central problem of modern liberal democracy.

Returning to the ancient world, one possible implication of Augustine's thought is that there is no necessary reason to trust *any* state, not even a Christian one: Even a state that claimed to rule in the name of Jesus Christ might turn out to be corrupt, impermanent, or both. Not only are earthly governments more numerous than divine governments, but impostors may abound, and they may not be recognized until it is too late.

It's perhaps understandable that Augustine did not dwell on this possibility. Notoriously, Saint Paul had taught Christians, and even slaves, to submit to their earthly rulers. The New Testament was never meant as a manual for political revolution, and some passages within it display a political quietism that to our minds may appear downright troubling.[14] As we shall see, these same passages were used for centuries as justifications for absolute political authority. And worse, by Augustine's time, Christianity was already deeply entwined in imperial power politics. Beginning with Constantine (272–337) Christian political leaders had begun defending their own place at the head of the empire. This happened first in the East, which still stood, and then later in the West, when self-styled emperors would claim to be two things that Augustine suspected were incompatible: They were, they said, both holy *and* Roman.

This put the Church in a difficult position. While imperial favor certainly beat persecution, the emperors' interventions were not always welcome. Still less welcome were the interventions of various unruly kings in the West, who were often tempted—like Charlemagne—to claim imperial prerogatives.

Of course, the Church was even more reluctant to endorse any attempt at revolutionary political transformation. It thus had a fine line to walk. As the eminent legal historian Harold J. Berman put it, for the early medieval Church, the city of man

> was without hope of redemption: it could only be abandoned for the realm of the spirit. St. Augustine and the church, generally, in the first ten centuries, were against revolutionary millenarian movements ... which tried to transform the social and political and economic realities of the here and now into a heavenly kingdom of the spirit. The rebirth of the individual Christian believer as well as the regeneration of mankind were understood to refer only to the eternal soul.[15]

As a result, Christianity was surprisingly slow to reform western Europe's autochthonous legal traditions. Even as it first conquered Rome and then spread throughout the West, Christianity brought little in the way of immediate, fundamental change to the states it found there. Again, following Berman, "the monasteries that sprang up all over Europe between the sixth and tenth centuries ... offered no program of secular law reform; they offered, instead, an ascetic life of work and prayer."[16]

The changes that Christianity did work were essentially twofold. First, Christianity's belief in the equality of all people before God tended to ameliorate—gradually, and certainly not completely—the status of women, slaves, the poor, and the helpless.[17] These changes were by no means a revolution, but they were at least a modest improvement. Ancient pagan historians and philosophers were much less widely read during this time, and many of their comparatively elitist theories enjoyed a diminished currency. In general, early medieval authors knew relatively little of either Aristotle or Plato. But by way of Cicero, who remained well known, medieval thinkers stayed in dialogue with Stoic philosophy, and that school of thought worked to much the same end. As the intellectual historian Ernst Cassirer has written, "The Stoic maxim of the fundamental equality of men was generally and easily accepted and it became one of the cardinal points" of medieval political theory.[18] As well it should, because it was so consonant with the message found in the Gospels.

Second, and much more troublingly, Christianity collaborated in the transformation of the ruler from a tribal chief (*dux*) into a king (*rex*). Once converted to Christianity, the king no longer represented only the deities of his tribe: he represented, at least potentially, a universal deity whose authority extended to many if not yet all people, regardless of tribe. He became, or aspired to be, the head of an empire.[19] The purposes of his government would necessarily change, and not clearly for the better.

Still, the conversion of pagan chiefs into Christian kings was not a development the Church could afford to criticize. A compromise of sorts was reached: Rule if you must, the Church effectively told both kings and emperors, but do it our way, and with respect for our independent—and perhaps prevailing—authority. This last became increasingly important during the high middle ages, during which the legal reasoning found in canon law developed as an independent source of legitimacy. This development, I will argue, is highly significant in the history of government's purpose.

As we have just seen, medieval political life was dominated by a complex conflict between church and state, one whose boundaries were drawn very differently from those of the present. Both church and state sought to credibly advance themselves as the truest defender of the Christian religion. Even "secular" states—a term first used in this era—were, by our lights, remarkably religious in their concerns, and this brought them into a deep, centuries-long conflict with the papacy. Increasingly, the struggle between them played out in the relatively new discipline of *law*.

Written law was ancient, of course, both in its constitutional form and in the form of specific injunctions and commands from the sovereign to individuals or groups. What changed most dramatically, during a period centered on the twelfth century, was the place of law in society, and its emergence as a distinct field of intellectual inquiry. In this era, the first law schools were founded. Scholars also wrote the first new law codes since the ancient world, in the form of Church, or canon, law. For the first time, the study of the law was understood as a field distinct from general scholarship. Literate legal scholars began to assert that their approach to the law was superior to the older, frequently oral and traditional understandings of law. Legal historian Harold Berman has characterized these changes, which endure to the present, as "revolutionary," and I think his assessment is correct.

All of these developments gave birth to a new idea, namely that the law was a body of reasoning, that it had an internal logic to it, and that the law, much like an organism, developed and grew over the course of time in response to challenges and opportunities.

Understood in this new way, law also became a thing that government might be *for*: On this view, governments existed to carry out what the law demanded. Law was not an aftereffect, a consequence, or an instrument of the state. On the contrary, the state was, or at its best could be, a servant of the law, and an outgrowth of the need for law.

Secular states still wrote laws, of course, and no one denied it. But if the body of the law considered as a whole really does have an internal logic independent of the state, then the possibility opens up that not everything a state does will necessarily count as a part of the law. It would not matter if the name "law" were given to it: Lawyers and legal scholars would (indeed must) decide whether that name was fairly applied. The members of the legal profession would make that determination guided by precedent, reason, and the existing, written body of the law, which they have dedicated their lives to studying and understanding.

Though little is known about his life, the twelfth-century canonist Gratian was one such scholar. His work has been important and influential enough that it deserves some analysis here. Nonetheless, he is seldom read today, apart from the few remaining specialists in canon law. This is understandable, as reading Gratian's Decretals can be a strange and disorienting experience. Early chapters begin reasonably enough; they define common legal terms, including "law," "ordinance," "canon," and "privilege." They also treat the reason why law exists.

For Gratian, following Isidore of Seville, "the purpose of legal enactments is to check human temerity and the capacity to do harm."[20] This to the modern mind likely sounds plausible enough. Again following Isidore, Gratian agrees that an ordinance should be "proper, just, possible, in accord with nature, in accord with the custom of the country, suitable to the place and time, necessary, useful, clear enough so that it contain no hidden deception, and not accommodated to some private individual, but composed for the common utility of the citizens."[21]

Again, so far, so good. Each of these descriptors seems at least plausible today. Each is at least theoretically also a reason why a jurist might choose to reject a purported law as being in no sense part of the law properly understood. Several of these qualifications even survive down to the present in recognizable form in the common law tradition. What surprises is not our seeing them, but our seeing them so early.

There then follows, as if from nowhere, a section on pregnant and menstruating women. Yet Gratian had a reason for his ordering: A widespread custom of his time, and one that remained in many parts of Europe for centuries, held that menstruating women and those who were either pregnant or recently delivered of a child were forbidden to enter a church. The Old Testament even seemed to recommend a similar custom, as Gratian noted. Informed by scripture, Gratian's readers may have been inclined to shrug at then-current laws and customs that forbade ritually impure women from attending church.

But this law served for Gratian as a model of a *bad law*, of a law that nature and reason rebelled against, and one that had no part in the law properly understood. Its placement early in the text would allow readers to see the sort of reasoning that Gratian believed was important to a good legal system.

"Natural law receives first place among all others .... For it began with the appearance of rational creatures and does not change over time, but remains immutable," wrote Gratian.[22] Even some divinely ordained laws—even some laws that came directly from God—were a part, in effect, of custom, and not of immutable reason or of immutable natural law. Just as the dietary laws had been explicitly overturned in the New Testament, so too the laws of ritual impurity should be no more. They existed only for a time and a place, and they were not to bind for all time. The ritual law was not eternal, and it was not suited any more to time or place.

Law was a nested hierarchy. A broad, overarching law, of a sort closely identified with the Divine, governed all. But within it, other laws could and should exist, laws that were responsive to time and place, to custom and convention. God could and did work within this type of law as well, and He had done so in laying out the ritual law of the Old Testament.

But law was not—or at least, it should not be—an arbitrary accretion of commands and prohibitions. When it did appear as such, or when it clearly was such, then reason should act either to reconcile conflicting laws or to choose among them the one that most accorded with itself. Implicit in the process was the idea that progressive, iterated reconciliations of laws and customs, the working of which was guided by reason, could produce better and better law. Conflict of laws was taken not as a defect, but as a positive virtue of the system: The presence of conflict was an opportunity for improvement; the ability to recognize conflict, and to apply reason to the problems it presented, was the work of law—and thus, in part, it was the work of government. Gratian himself titled his work *Concordia Discordantium Canonum*, or *The Concordance of Discordant Canons*. As Katherine Christensen writes, what Gratian did was to

> present his sources, one by one, and when ... contradiction appeared, he inserted his own commentary, in sections referred to as dicta ... . in which he suggested how the authorities might be reconciled, or how one was to be held preferable to others: how, ultimately, the discord could be resolved ... . General councils outranked their regional counterparts; popes outranked bishops ... . An opinion on some specific and limited matter, from however eminent a source, might not deserve the status of a general rule, needing to be applied in all or in greatly different circumstances.[23]

Much of this will be startlingly familiar to a modern lawyer accustomed to the common law tradition. It will likewise be familiar to an attentive reader of F.A. Hayek's *Law, Legislation, and Liberty*, which commends a similar method of rulemaking. The clerical context may be unfamiliar, but the methodology nonetheless unfolds along similar lines.

One significant difference exists between Gratian's account of the reconciliation of conflicting laws and the accounts given by more modern legal theorists: It now seems strange to us that, in Gratian's paradigm case at least, God was apparently *on both sides* of the conflict. Qua establisher of the ritual law, God excluded the ritually impure from worship. Qua establisher of reason, God urged all to attend worship, without exception. In determining that the ritual law must answer to reason, although both were certainly authored by God, the legal process seemingly reconciled God to Himself.

We might take it as a given that when a conflict of laws exists, right reason lies only on one side, while error lies on the other. And if we believe in a God, we might want to identify him closely indeed with right reason, and never with error. Even if we don't believe, the more usual understanding is that reason is unitary and does not conflict with itself.

Yet we should be careful in applying too freely the considerations of abstract philosophy to legal theory. It is frequently the case that law is *not* a matter of right reason alone, but also a matter of convention. In many cases, the really important task at hand is simply achieving coordination on one ethically neutral convention or another. In such cases, many different rules may more or less equally suit the purpose, provided only that a single one of them is chosen, and that the others are abandoned. Contradiction in these cases does not indicate error on either side. It simply indicates a suboptimal outcome. Here, the controlling criterion is not right reason, but a uniformity of convention. God, we might say, is on the side of convention, but not on the side of two or more conflicting but otherwise neutral conventions at once.

There are, then, potentially two types of conflict in law: A law may conflict with right reason, or it may conflict with another merely conventional law. Jurists must rule in either case. Determining whether a conflict of laws is one of justice or of convention is a task for philosophers, which in some sense our jurists must also be. Performing either of these reconciliations, Gratian suggested, is a legitimate task of government. The ritual laws that Christians had largely abandoned would appear, to him, to be something like a convention, albeit one whose time had come and gone.

Problems, of course, remain. One in particular was the establishment of a *natural* foundation for purportedly natural law, one whose articulation was at least convincing enough to command moral authority, and, when necessary, to overturn defective conventional law. "[D]ivine ordinance is indeed consonant with nature … . So both ecclesiastical and secular enactments are to be rejected entirely if they are contrary to natural law," wrote Gratian.[24] But what exactly *was* natural law?

To put it mildly, this has been a difficult question in the history of western philosophy. Gratian himself was altogether vague on the matter. As the eminent medievalist Stanley Chodorow has written,

> [The meaning of natural law] was a problem that interested and perhaps even amused the earliest commentators on the Decretum. Stimulated by Gratian's apparent inability to give a single, comprehensive definition for *ius naturale*, the later canonists exercised their imaginations in order to list all the possible meanings of the term. One canonist found six such meanings.

Another limited himself to just five.[25] Whatever the case may have been, it was surely better that *some* limitation existed in principle to the power of organized violence, present in both the state and the Church. If defects remained, they might be sorted out later, through the iterated and reiterated reconciliation of customs and canons. Perhaps time would also reveal the nature of natural law.

Such was the hope, at any rate, and we ought not to dismiss the significant advance that such reasoning presented in the struggle to define what government was for. The emergence of this new understanding of law promised for the first time to give a purpose to the state that was both *defensible* and *limited in scope.*

And it was to bear fruit later in history, notably in the works of Thomas Aquinas (1225–1274). Aquinas synthesized Aristotelian philosophy with Christianity; in doing so, it is clear that he remained at least somewhat sensitive to the early Christian idea that government might be an affliction. Aquinas argued that war—the state action *par excellence*—could only be considered just under a set of very exacting circumstances: Wars could not be waged for earthly wealth or power; they could not be indulged in by just anyone, but only legitimate authorities; and wars should always be conducted for the sake of, and with a view toward, reestablishment of peace. Conduct within war was likewise to be constrained, so as to minimize collateral damage and cruelty. By Aquinas' account of just war, nearly all wars were unjust, both in their ends and in their means.

All of this constitutes a significant improvement, by our understanding, over the ancient conception of the state and its purposes. Recall that for the ancient historians, the state existed as a sort of theater of virtue; the virtuous man acted in ways that the state laid open to him, and his success in battle or in statecraft demonstrated his (earthly, martial) virtue. That others might suffer in the process, and that their suffering might be morally significant, was in ancient thought decidedly the minority opinion. Indeed, the ancient consensus view of the state did not and could not impose any clear limits on what the state could do. The medieval one just possibly did, and for a variety of reasons, many of which remain compelling today.

As we have discussed, it is easy to see how Stoicism and Christianity had much in common. Medieval writers certainly knew both rich and poor, both powerful and powerless, and yet they firmly believed that man was created in the image of God. All mankind, regardless of social station, thus shared a common origin. Christians and Stoics also both held that all men were mortal, and that all arrived at a common end. For both, our mortality was suffused with philosophical implications. Christians averred that all would be judged by God, whose judgments would place no value whatsoever on earthly wealth or power; these, for the good Christian, were at best responsibilities to be borne carefully—and at worst, they were distractions or impediments.

Stoicism similarly disdained material possessions and political power. It argued that all men, whether Greek or barbarian, slave or free, citizen or alien, were to be judged by a common moral standard. Christianity is in many respects a cosmopolitan philosophy, and in Stoicism it found a startling resemblance to itself. From these common traits came at least the potential for a similar ideology of governmental restraint.

The advent of the law as a living, developing body of reasoning also fit well within Christianity. It gave the state a purpose that did not resolve simply to the morally dubious aims of personal advantage or personal glory. The law might even be brought around to validating the equal moral worth and dignity of all individuals, of people who were all created in the image of God, and who were all worthy of a fair hearing. One very long-term project of the state—one might even say a project that would ultimately be realized in Heaven—was thus to advance society in the direction of the reasoning of the law, which was after all nothing more than God's reasoning about his own creations.

Because we are mere mortals, it will be a plodding, error-prone process. As the intellectual historian J.G.A. Pocock has written in a different context, "In a fallen world, even divinely commanded authority has the character of praxis rather than of pure form."[26] Where subjects acted against the law, the state was to punish. But where the law was defective, the state was tasked with amending it—and thus, with amending itself. The state existed, then, to discover and to articulate a particular kind of reasoning, and to achieve in practice what that reasoning demanded in theory.

Gratian and his successors were among the first to express the idea that, at least for now, in this fallen world, one of the key purposes of government was ultimately one of self-discovery: Government had to explore, discuss, and debate its own purposes, because *these purposes might not be fully understood yet.* Beyond the mere reconciliation of conventions, we may not yet know exactly what justice consists of, even while we clearly have *some* idea of it. The continuing conceptual refinement of the law, guided by principles whose particular applications are not necessarily clear to anyone at the time of their enunciation is perhaps a risky strategy, true, but no strategy of interpretation is risk-free, and this one at least promises peaceful resolution of contradictions over time.[27]

This modest epistemological stance likewise entailed modest government. Many state actions simply could not be justified with reference to existing law. Others, when their principles were articulated, appeared intolerable: The high middle ages saw both the demise of trial by ordeal and the advent of written trial records, so that others could later review whether proper procedures had been followed. Always, the possibility of error was admitted.

Now, the study of constitutional law had of course existed in the ancient world, but the legal revolution of the high middle ages was distinct from it in one crucial way: The earlier scholarship had little binding force against state actors. That was to change in the medieval era, when the very people doing the scholarship were also the ones actively running the system of justice, as part of an elite class whose purpose, and whose allegiance, appears to have been to the institution of law, rather than to any particular ruler, or to the all-too-nebulous polity as a whole, or to personal gain more narrowly understood. Jurists were committed to God, to the law and—to be sure—to their own professional interests. The key factor here was that not one of these commitments aligned them consistently with other contenders for power.

Jurists' censure was not a wholly academic affair, then. When lawyers and judges oppose state actions strenuously enough, the state loses the benefits of their service, which are considerable. The increasingly complex state apparatuses of the western kings, the Holy Roman Emperor, and the papacy had all come to depend upon such services by the twelfth century. Keeping this professional class happy was important to a well-run state, and keeping them happy meant, in part, at least a rough compliance with the rule of law as the jurists understood it.

This is not to say that the presence in a society of the idea of the rule of law guarantees that all state actions will be just. Nor does it guarantee that the state's powers will always be limited in practice. On the contrary, western Europe in the twelfth century witnessed not only the development of a nearly modern conception of law, but also the development of what R.I. Moore has called the "persecuting society," in which heretics, lepers, Jews, and other outsiders were subject to much more systematic persecution than they had previously experienced.[28]

Disturbingly, the two events are by no means unrelated. If Moore is right, the persecuting society—which put Jews in ghettos, burned heretics and exiled and vilified lepers–was, like the rule of law, the product of the newly ascendant legal profession. The clerics who made up this profession were literate and adept in working the machinery of justice. They were therefore quite useful to the state, but they brought to the state an agenda of their own.

It is understandable, if not excusable, that they jealously guarded their newfound privileges against even some fairly unlikely challengers: Heretics stood to undermine the Christian religious order, on which their profession depended. Jews were literate and yet obstinately not Christian. Even lepers had, for a time at least, passed for holy men in the West, and their divine affliction might serve as a rebuke to the legal profession. The persecution of these three groups, Moore argues, was partly a consequence of the rise of the lawyer in the West. As was the advent of the rule of law. It would take centuries of cruelty to untangle what we only with hindsight see as two conflicting threads: impartial law, attendant only to its own logic, and what we would have to call the opposite, partial law, which is no law at all.

One final possibility must be mentioned before leaving the medieval world, even if it does not appear to have been defended often, or ever, at the time: It might prove that the purposes and the logic inherent in the elaboration of the law are of a type that the state is simply not very good at carrying them out. That is, pursuing the logic of the law in a consistent fashion might ultimately condemn not just particular state actions or particular enactments as undeserving. The logic of the law might go further, and in time it might ultimately prove that the law's purposes are better served without a state than with one.

Again, no one at the time seems to have believed this or to have considered it. Still, it was a tremendous advance simply to suggest that the state was answerable to an independent moral order, that it existed to realize that order, and that in practice the state will frequently—even regularly—fail at doing so. It was a further advance to suppose that the domain of state action was properly contestable, and that it should be subject to a process of orderly and public dispute. The implications of these insights were vast, and they remain so today.

Gratian himself did not see the majority of these implications, and of course we should not expect him to have done so; still less should we expect him, if risen from the dead, to give his blessing to the Anglo-American common law tradition as we might find it in any particular later era, including our own. The virtue of a common law system, and of the reconciliatory approach to law more generally speaking, is that it allows a process of real and yet orderly improvement in law and custom in the midst of continuity. It does not supply trans-historical consensus, which would clearly be unreasonable to demand of any legal philosophy at all.

Notes

1. Augustine of Hippo, *City of God*, Book I, Preface, available at http://www.newadvent.org/fathers/120101.htm.
2. Ibid., Book II, ch 21, available at http://www.newadvent.org/fathers/120102.htm.
3. Ibid., Book II, ch 21, available at http://www.newadvent.org/fathers/120102.htm.
4. Ibid., Book III, ch 10, available at http://www.newadvent.org/fathers/120103.htm.
5. Ibid., Book IV, ch 4, available at http://www.newadvent.org/fathers/120104.htm.
6. Cicero, *Political Works.* http://oll.libertyfund.org/titles/546#Cicero_0044-01_534.
7. Buchanan, "Public Choice."
8. Olson, *Power and Prosperity*, pp. 6–9.
9. Augustine of Hippo, *City of God*, Book V, ch 21, available at http://www.newadvent.org/fathers/120102.htm.
10. Levy, "On Taking Politics Less Seriously," https://www.youtube.com/watch?v=cyRUJtfRfK8, at 30:04.
11. Augustine of Hippo, *City of God*, Book II, ch 19, available at http://www.newadvent.org/fathers/120102.htm.
12. Jacob T. Levy, "On Taking Politics Less Seriously," https://www.youtube.com/watch?v=cyRUJtfRfK8, at 18:40.
13. Tolstoy, *The Kingdom of God Is Within You.*
14. Ephesians 6:5: "Slaves, obey your earthly masters with respect and fear, and with sincerity of heart, just as you would obey Christ." 1 Timothy 6:1: "All who are under the yoke of slavery should consider their masters worthy of full respect, so that God's name and our teaching may not be slandered." Explanations for these passages have been provided, but the fact that they cry out for an explanation only proves the point.
15. Berman, *Law and Revolution*, p. 27.
16. Ibid., p. 64.
17. Ibid., p. 65.
18. Cassirer, *The Myth of the State*, p. 103.
19. Berman, *Law and Revolution*, p. 66.
20. Gratian. *Treatise on Laws*, p. 12. D4.C1.
21. Ibid., pp. 12–13. D4.C2.

22. Ibid., p. 16. D5.
23. Ibid., pp. xv–xvi.
24. Ibid., p. 32. D9.C11.
25. Chodorow, *Christian Political Theory and Church Politics*, pp. 98–99.
26. Pocock, *The Machiavellian Moment*, p. 353.
27. Sandefur, "Liberal Originalism," p. 489. Available at SSRN: http://ssrn.com/abstract=656136.
28. Moore, *The Formation of a Persecuting Society*.

CHAPTER 5

The March of the State in the Early Modern World

Contrary to popular belief, the Renaissance was not a propitious time for individual liberty. Thinkers of the sixteenth and seventeenth centuries recast the state, and the king in particular, as the agents of God on earth. This was to have exactly the effects on political thinking that one might have expected.

The king (or the state) now took a high place in what was termed the Great Chain of Being: This was an ontological and moral order, pervading the entire universe, purportedly established by God, and stretching from Him through angels, human societies, animals, plants, and minerals. Every person was said to have a place, and wisdom consisted, first, of knowing one's place and, second, of remaining within it. Although such rank orderings were common in the medieval era and even in Platonic philosophy, the Renaissance came to apply them to government with a particular fervor and sincerity, and to insist on a particularly important place for the agents of the state within the moral order: The state and its agents were ordained directly by God, and the king answered to Him alone.

The most illustrative figures in this story are to my mind Niccolò Machiavelli (1469–1527), Tommaso Campanella (1568–1639), and Jacques-Bénigne Bossuet (1627–1704). I will discuss each in turn. Machiavelli, I will argue, was significantly Platonist despite himself. Campanella was knowingly and eagerly Platonist. Bossuet was perhaps history's greatest exponent of absolute monarchy, and he had thoroughly absorbed the notion of the Great Chain of Being; I will suggest that he

© The Author(s) 2017
J. Kuznicki, *Technology and the End of Authority*,
DOI 10.1007/978-3-319-48692-5_5

made a stronger, more influential, and more coherent case for absolutism than either Robert Filmer or Thomas Hobbes, the two most familiar absolutists in the Anglophone world.

From Bossuet, we will turn to the social contractarians, beginning with Hobbes; although Hobbes falls earlier in the chronology than Bossuet, he represents a distinct break from the thinking of these other authoritarians, and he should not be classed among them.

Machiavelli: The Return of Ancient Virtue

The Florentine diplomat Niccolò Machiavelli is often described as a pagan thinker, or, following an older and less accurate tradition, as an atheist. Neither of these terms is completely wrong, but both can be misleading. If a *Christian* is one who places the Christian idea of God at the highest point in his ontology and his morals, then a *pagan* might be said to place either the pagan gods or some pagan philosophy in a similar place. An *atheist* would deny both pagan and Christian deities in favor of some other end.

For Machiavelli, that highest end was the state. It is thus not wrong to term him an atheist, but the most accurate term for him would be *statist*. That term names what was for him the greatest good: To Machiavelli, the good in anything was to be found in its usefulness to the state. Many things, including human individuals, were *made for* the state. Some other things, including honesty and deception, kindness and cruelty, and religion itself, could and should be judged as either good or bad instances of their type depending on whether they were good or bad for the state: A good honesty would serve the state, while a bad honesty would impede or embarrass it. A good cruelty would benefit the state, while a bad cruelty was only bad because it hindered the state.

Importantly, Machiavelli deployed a word in this context that is cognate to our own (*stato*). This is not to say that his idea of the state is precisely like our own, or that every observation of Machiavelli's might apply equally well in our own day. It does mean, however, that the vocabulary of political theorists in this era is growing similar to our own. Renaissance polities are likewise a good deal more like our own when contrasted to ancient polities, and the cross-cultural and translation difficulties are accordingly much smaller.[1] With Machiavelli, readers can recognize with greater ease certain ideas about the state that they have probably already encountered in contemporary life.

For example, Machiavelli held that several types or classes of men existed; those of the lower class he judged to be good insofar as they were "poor, militarized, honest, and obedient," as Isaiah Berlin put it.[2] They were to serve the state in war, and to stand ready for war even in peacetime, transmitting martial virtue down through the generations, insofar as that was possible. They were to avoid the temptations of luxury and courtierism.

This view of the citizenry is not without its present-day appeal, and it can be found as a component of many varieties of nationalism today. Indeed, Machiavelli's praise of the honest citizen-soldier was to spawn a vast literature and an entire school of thought constructed along similar lines, albeit avoiding Machiavelli's embrace of amoralism and of the state as the highest good. That school is known as classical republicanism. Many great and undoubtedly sincere defenders of individual liberty, including Algernon Sidney and James Harrington, should also be counted among the classical republicans. They were inspirations to the American Revolution and are arguably why the United States became a republic rather than yet another monarchy.

So much must be said in fairness to Machiavelli. Yet a naive reader would not come away from Machiavelli with the sense that this was a man who cared much for either individual liberty or limited government. And the naive reader would be more or less correct: True, interpretations of *The Prince* do exist that hold this work to have been satire, and it may even have been one. But Machiavelli's *Discourses on Livy* is certainly not a satire, and it shares a common thread with *The Prince*; that thread is that the state must be considered the highest good, regardless of its institutional structure. In my view, this commonality makes the satirical reading of *The Prince* doubtful in many respects.

In both works, we find in addition to the poor, self-sacrificing citizenry a higher class of men, who governed by right. To Machiavelli, proper statesmen possessed an elusive attribute, *virtù*, which set them apart from others. *Virtù* did not mean virtue in the ethical sense, which denotes a settled habit or disposition toward the good. On the contrary, *virtù* meant a sort of dynamic but amoral power, a capacity to strive mightily and to take decisive action, whether or not an ethicist might approve of it. (One component of *virtù*, it might even be said, was a complete lack of squeamishness about what ethicists might think of one's actions.)

In all, *virtù* meant something close to the word "manliness" as it might be used by a male chauvinist; indeed, the word's Latin root, *vir*, simply means "man." *Virtù*'s antonym was not vice, but softness or effeminacy.

Virtù was both a title to rightful rulership and the means by which one attained and exercised it. A prince might possess *virtù*, but so might a whole citizen body, at least at times. The fact that *virtù* might be found in either a republic or a principality explains how Machiavelli might have written—without any real contradiction—both *The Prince*, which is a manual for absolute rule, and the *Discourses*, which is a treatise on how to run a classical republic. Individuals who were not princes or the citizens of republics were unlikely to have opportunities to exercise *virtù*, as they were generally denied political agency of any sort. It was not unthinkable, of course, that a peasant or a slave might rise to great political power. It was merely so unlikely that such events did not merit much study.

One might also liken Machiavelli's view of the state to that of Plato. Although Plato was an idealist, and although Machiavelli passes for a realist, the two arrived at some strikingly similar descriptions of the state, and at some nearly identical prescriptions for running it, and for the state's relations to the rest of society.

Recall that for Plato, a true statesman could and should make laws on a case-by-case basis, and that this was not a power properly entrusted to anyone else. For all but the statesman, a written law was to prevail, because true statesmen were rare. This special distinction is similar to the one reserved for the Machiavellian possessor of *virtù*, who is permitted, when led by *virtù,* to act in defiance of custom, law, and morals—provided only that his acts serve the state.

As an example of the exceptional privilege given to political adepts in both Plato and Machiavelli, let us consider the nearly identical ways that each may treat religion: Both Plato and Machiavelli held that whenever possible, religions should be tailored to serve the good of the state and its rulers. Plato proposed to found a state on a frankly false origin story, one shrouded in an equally false religion. This religion would establish a permanent guardian class and would convince all social orders to remain forever in their places. In this way, Plato claimed, the state would be made more permanent. Permanence was to be desired because it was an attribute of the good.[3]

Machiavelli wanted exactly the same. His *Discourses* takes the form of a lengthy commentary on Livy's *History of Rome*. There Livy had recounted the legend of Romulus, the first king of Rome, and the (perhaps slightly more historical) legend of his successor, Numa Pompilius. Machiavelli showered Numa with revealing praise: "Numa, finding the people ferocious and desiring to reduce them to civic obedience by means of the arts of peace, turned to religion as the instrument necessary above all others for the maintenance of a civilized state," Machiavelli wrote.[4]

Truthfulness was not among this religion's virtues. Machiavelli conceded, or perhaps boasted, that Numa only "pretended to have private conferences with a nymph who advised him."[5] The result was "a religion that facilitated whatever enterprise the senate and the great men of Rome designed to undertake."[6] And thus, *although it was both false and expounded insincerely*, Roman paganism nonetheless deserved praise. Machiavelli was not so much a sincere pagan as he was an advocate of paganism considered as a noble lie. Christianity, meanwhile, might and often did frustrate state power, and for that it was to be blamed.

Yet even in ancient Rome, the wheels eventually came off the cart. As Machiavelli wrote:

[W]hen the oracles began to say what was pleasing to the powerful, and this deception was discovered by the people, they became incredulous and inclined to subvert any good institution.

The rulers of a republic or of a kingdom, therefore, should uphold the basic principles of the religion which they practise in, and, if this be done, it will be easy for them to keep their commonwealth religious, and, in consequence, good and united. They should also foster and encourage everything likely to be of help to this end, even though they be convinced that it is quite fallacious.[7]

The noble lie only worked when the people believed in it; religion only did its "good" work—good, that is, for the state—as long as kept its power to dupe. And Machiavelli favored the pagan religion not because it was true, but because, although false, it was *useful*.

By contrast, the Christian religion was a scandal and an embarrassment to the state. Nowhere was this clearer than in Machiavelli's Italy, which the Church kept divided into squabbling city-states:

It is the Church that has kept, and keeps, Italy divided ... For, though the Church has its headquarters in Italy and has temporal power, neither its power nor its virtue has been sufficiently great for it to be able to usurp power in Italy and become its leader; nor yet, on the other hand, has it been so weak that it could not, when afraid of losing its dominion ... call upon one of the powers to defend it against an Italian state that had become too powerful.[8]

There is not a single good word in this passage for the doctrine of the Church, or for its morals. Nor are there any bad words for them either. But there is much condemnation for the Church's lack of efficacy or power—its lack of *virtù*, in the Machiavellian sense. Had the papacy conquered and held all of Italy, Machiavelli might even have praised it. Though not, perhaps, for reasons that would have been pleasing to the Church.

As we have discussed, Plato's statesman—the lawgiver who possessed an unfailingly certain system of knowledge, who could institute an eternal, unchanging order, if only we were good enough to obey him—was decidedly a skyhook. Machiavelli would have agreed on this point, I think. But to Machiavelli the man of sufficient *virtù* could still command obedience from his fellows and could still turn the course of history. Such men were not necessarily perfect lawgivers, as Plato might have had it, but they were rare and precious, and their leadership abilities—their *virtù*—meant that they could overcome the formless hodgepodge that was the stuff of ordinary history.

We are close once again to the ancient idea that the state was a theater of virtue, and it is now easily understandable why Machiavelli is so often termed a pagan. How, though, did one recognize *virtù*? Most obviously, to see it was to know it: One judged *virtù* by the fact of its exercise, and by observing its practical effects in the world. Machiavelli claimed in particular that his patron, Cesare Borgia, was a man of great *virtù*, and he often had Borgia in mind when discussing the concept.

Machiavelli's science of government, then, the body of knowledge upon which he based political authority, was history, and particularly the history of power politics and great men. As he explained in *The Prince*, "[A] prince should read history and reflect on the actions of great men ... [taking] as a model for his conduct some great historical figure who achieved the highest praise and glory by constantly holding before himself the deeds and achievements of a predecessor."[9] And in the *Discourses*, "[E]verything that happens in the world at any time has a genuine resemblance to what happened in ancient times ... [T]he agents who bring such things about are men, and ... men have, and always have had, the same passions, whence it necessarily comes about that the same effects are produced."[10]

For Machiavelli, history stood in a position analogous to numerology in Plato's political thought: It served as a purportedly justificatory science. Given our contemporary evaluation of these two bodies of knowledge, Machiavelli comes across as having identified the more reasonable foundation. Yet what Machiavelli thinks he can discern from history is more than most historians now believe that it can provide. It remains unclear that *virtù* can be either learned from history or reduced to the scrupulous applications of its lessons. It is even open to dispute whether such a trait exists at all, except in the trivial sense that some statesmen seem to fare better than others. As David Hume remarked about Machiavelli, "We have not as yet had experience of three thousand years; so that not only the art of reasoning is still imperfect in this science, as in all others, but we even want sufficient materials upon which we can reason." As a result, he found Machiavelli's conclusions "extremely defective."[11] History may be the intellectual grounding for a science of government, but history itself is not yet sufficiently developed. Nor will it be, if Hume was right, in any of our lifetimes.

We should recall at this point, however, that what is demanded of a science of government in this mode of thinking is not, in fact, the truth, nor even the earnest search for truth. What is needed is simply the power of persuasion directed at a suitable audience. As Plato had hoped of his numerology, Machiavelli likewise hoped of his humanistic and not entirely fruitless study of history. He did not however seek knowledge for its own sake. History was the science of authority.

What, though, does all of this have to do with our own central question—what is government *for?* Our question seemingly presupposes that government, even within the institution of the state, exists *for a reason or reasons external to itself,* and that these reasons can be articulated and perhaps even refined over time. Machiavelli was uninterested in such questions, not because he was uninterested in the good, but because he identified the highest good with the state itself. It might naturally cause other goods to flow to the governed people, but these were certainly not the reasons for having a state.

That's because when one writes of the highest good—the thing that is good in itself, and good for its own sake—one does not ask what this highest good is good *for:* If it were good *for* some other thing, then *that* thing would be the highest good. The devout Christian, for example, does not properly ask what God is good for, or how one might use God. These questions would be impious, because God is held to be the supreme good. *That,* or something very like it, is the place that the state occupied for Machiavelli. Though it is perishable, though it is a human creation, though it is obviously flawed in all sorts of ways, the state is still the highest end.

As the intellectual historian Ernst Cassirer has written, "The sharp knife of Machiavelli's thought has cut off all the threads by which in former generations the state was fastened to the organic whole of human existence. The political world has lost its connection not only with religion or metaphysics but also with all the other forms of man's ethical and cultural life."[12] To one who accepted Machiavelli's principles, no force existed that could call the state to account; if all went well, there would *never* be a day of reckoning for any crime committed in its name.

Cassirer did not claim, and nor would I, that Machiavelli grasped the full implications of his doctrines. But Cassirer did claim that these ideas were "pregnant with the most dangerous consequences," which seems quite correct. In the medieval world, conversations about the nature and extent of state power were common, and a part of the state's legitimacy seemed to come from its capacity to entertain the prospect that it might serve higher ends, and that, judged by these ends, the state might even be wrong. This is not to say that the medieval state was a defanged beast, or that the ways of suffering or cruelty were unknown to it. But it is hard to escape the conclusion that the medieval world possessed at least some of the intellectual tools necessary to think its way out of state omnipotence, and that the Renaissance struggled mightily to unmake them.

CAMPANELLA: THE ASTROLOGICALLY PERFECT STATE

Loosely following Karl Popper, we might say that there are two types of theories of government. One type holds that a special science or knowledge of government is—or soon will be—capable of arresting the progress of time, of creating an eternal, unchanging society, an order that will constitute perfection in itself. Perhaps, as some would have it, this perfect state will prepare the way for the Kingdom of God.[13] The other type of theory holds that Fortune and change are unavoidable, and that it is not for us, or for anyone else, to set up an eternal city. We can't possibly know everything that we would need to do it. The best we can do is to learn from history, and to make note of how various experiments have gone, not in the hopes of constructing a final theory, but simply with the goal of avoiding past mistakes and improving our states the best we can. Machiavelli belonged to the latter camp, even if his purposes for government were expansive and overbearing. Tommaso Campanella, however, belonged to the former. Like Plato before him and Marx afterward, Campanella looked forward to the end of history, in the form of a perfected and permanent society, one whose fundamentals would no longer be subject to Fortune. Campanella lived and worked roughly a century after Machiavelli, and in the intervening years, the influence of Platonic philosophy had grown enormously. Machiavelli had built his authority primarily on the discipline of history, and where resemblances between Machiavelli's and Plato's theories of government are not necessarily obvious, Campanella's Platonism is impossible to miss: Like Plato, he also claimed to possess an exact technology of government.

For Campanella that technology was, improbably enough, applied astrology. Campanella's apparently earnest belief in the predictive value of the stars can make his most well-known work, *The City of the Sun*, difficult to take seriously. But I believe that we should do so all the same; although political theorists seldom consider Campanella, he is nonetheless a key founder of political utopianism, which has been tremendously influential.[14]

Taking Campanella seriously does not require us to take his astrology seriously. The claims of astrology are uninteresting and uninformative. But the powerful urge to believe in a comprehensive technology of human governance ought to command our utmost attention. As Campanella shows, this impulse can profoundly inspire us, and it can even inspire us toward absurdities.

In *The City of the Sun*, Campanella presents us an imaginary city, Solaria, that is run according to his principles. It is a clockwork city, a scientifically planned utopia that serves as a monument to the nearly universal—and always dangerous—desire for comprehensive social planning. The city's inhabitants lacked just one thing, Campanella claimed, namely revealed religion. But in their natural philosophy, we are told that they lacked nothing. As with Plato's *Republic*, the planning began at the city's inception. Here are the conditions that supposedly obtained at Solaria's founding:

[T]he fixed signs at the four corners of the world—the sun in the ascendant in Leo; Jupiter in Leo oriental to the sun; Mercury and Venus in Cancer, but so close as to produce satellite influence; Mars in the ninth house in Aries looking out with benefic aspect upon the ascendant and the apheta; the moon in Taurus looking upon Mercury and Venus with benefic aspect, but not at right angles to the sun; Saturn entering the fourth house without casting a malefic aspect upon Mars and the sun; Fortune with the Head of Medusa almost in the tenth house ... Being in a benefic aspect of Virgo, in the triplicity of its apsis and illuminated by the moon, Mercury could not be harmful; but since their science is jovial and not beggarly, they were not concerned about Mercury's entering Virgo and the conjunction.[15]

The Solarians, it seems, were a patient people: I have been unable to determine when, if ever, the full and exceptionally rare set of conditions listed above may have occurred. With the help of some software I have ruled out the years 558–2091 CE.[16]

One modern commentator has suggested that the Solarians might simply have arranged a mechanical model of the planets, known as an armillary sphere, to *display* the appropriate conditions, the planets' actual positions notwithstanding.[17] If so, what the Solarians practiced was not astrology, per se, but a sort of technological magic, one whose performance would at least not be impossible. Whatever the case may have been, the citizens of Campanella's utopia "laugh at us because we are careful about the breeding of dogs and horses while we pay no attention to our own breeding."[18] In Solaria, state planning also reaches into the bedroom:

> Since both males and females, in the manner of the ancient Greeks, are completely naked when they engage in wrestling exercises, their teachers … can determine whose sexual organs may best be matched with whose. Consequently, every third night … the young people are paired off … The exact hour when this must be done is determined by the Astrologer and the Physician, who always endeavor to choose a time when Mercury and Venus are oriental to the Sun in a benefic house and are seen by Jupiter, Saturn, and Mars with benefic aspect. So too by the sun and the moon that are often aphetic. Most frequently they seek a time when Virgo is in the ascendant, but they take great care to see that Saturn and Mars are not in the angles, because all four angles, with oppositions and quadratures, are harmful; and from these springs the root of the vital power and of fate, which are dependent upon the harmony of the whole in relation to its parts.[19]

Again, this would have been an exceedingly rare set of astrological occurrences. It is difficult, then, to understand what may have been meant in this passage by "every third night": it would often take years for all of the listed conditions to be satisfied. Perhaps, as with Plato, this was all intended for show; perhaps it was a noble lie, and not a practical breeding program, but if so, it was a particularly bald faced lie. Or perhaps the Solarians again had recourse to a mechanical model of the heavens, which they adjusted in miniature to resemble the conditions that they required for astrologically correct breeding.

The stars ruled Solaria's geography as well as its husbandry. The city was laid out in "seven large circuits, named after the seven planets."[20] Each was walled for defense. At the center was a circular temple, on whose altar there could be found a terrestrial and a celestial globe. At the center both physically and authoritatively, Solaria's prince was called Sun, and he was assisted by three underlings named Power, Wisdom, and Love. "For every virtue that exists among us, they have an official," Campanella wrote with approval. Lower officials all reported up one of several chains of command, which reached their apex at Sun, who was sometimes also called the Metaphysician.[21]

The officer named Love oversaw the breeding program outlined above. Power controlled matters of war, while Wisdom took charge of the sciences and the arts.[22] Individuals' careers were determined in at birth: "the particular inclination of each person is seen in his birth, and in the division of labor no one is assigned to things that are destructive to his individuality but rather to things that preserve it."[23]

All the Solarians' property is held in common, as is typical in platonic utopias. They have little use for trade, and their children "laugh when they see the merchants willing to give away so many things in exchange for a bit of silver."[24]

The entire society was arrayed, and its duties were delineated, almost as one might find in a Renaissance catalog of knowledge; extensive knowledge claims were precisely what the Solarians boasted of, and what they founded their governing system upon. The fact that we consider astrology preposterous does not change the fact that Solaria should be classed as a technocracy, that is, a government that is legitimate by virtue of its supposed technical competence. As we have seen with Plato, such a vision would appear able to appeal even without any scientific evidence to back it up: It runs on credence, not on evidence, and in particular it runs on a form of credence that many seem willing to give.

As to this government's purpose, it would appear to be the alignment of human society with transcendent astrological forces. Following the determination of the propitious moment for various activities, there is—and can be—no legitimate debate about what to do. The stars have spoken, and dissent is at an end.

They would of course be remarkably fortunate stars, if they never produced an oversupply of physicians, or an undersupply of shepherds, and if they never needed correctives after the fact. One begins to wonder about the not at all anachronistic question of Fortune: Renaissance thinkers, Machiavelli among them, worried about the inherently unpredictable nature of human events, against which they often claimed to have no permanent remedy. This concern with instability brought even Machiavelli to temper and qualify many of his political recommendations. His prince was a risk-taker. As one who intervened in unpredictable human events, a prince could not be otherwise. And taking risks meant that sometimes he failed, as had Machiavelli's paradigm prince, Cesare Borgia. For Campanella's Solarians, failure never seems to have arrived, and no provisions are made for it.

The Renaissance humanists' ongoing concern with political and social instability over time—and the sharp contrast that social instability made with the perdurable truths of logic—offered fertile ground in later centuries for thinking about the institutions of government. Humanism eventually birthed a thorough and searching account of political corruption, institutional integrity, and eventually even commerce and credit.[25] For Machiavelli, writing early in the development of this tradition, Fortune offered opportunities for those with *virtù*; but it ruined those who lacked it. Uncertainty, though, was everywhere.

Now, historians know that Campanella was not ignorant of Machiavelli, so it may be worth asking: Did Campanella really believe that astrology (and communism) could insulate a polity from Fortune? Did he mean what he wrote? Or was he indulging a reverie? Given how he wrote *The City of the Sun* while imprisoned with scant hope of release, might his work even be a bitter satire, one aimed against optimists everywhere?

Modern commentators generally agree that Campanella was in earnest. He disdained Machiavelli for what he saw as the latter's Aristotelianism: Machiavelli insisted upon the importance history and contingency; Campanella thought that these were to be swept aside. As one modern scholar wrote, "On the very same grounds which Machiavelli would be appreciated by the modern age ... Campanella ... takes violent issue."[26] For Campanella, the study of politics considered as an earthly, fallible art was not a mark of intellectual maturity. On the contrary, it led to the sort of thinking that would use religion and morality merely as tools. In disdaining Machiavelli, Campanella set himself up as a true believer, and not as an (open?) proponent of a noble lie.

Yet the system Campanella propounded had, if anything, even less to do with liberty than Machiavelli's. The latter at least recognized freedom from foreign domination as a legitimate goal, and there is certainly something of liberty to that. In Campanella's thinking, meanwhile, religion was natural—thus not the product of a noble lie—but by the same token, religion was to aspire to universality, and it was to do so using every possible means. Even the Solarians themselves, so perfect—and so perfectly controlled—were to yield to Christian religious authority when it arrived. As if, one might say, the Christian religious authorities of Europe were doing any better in governing themselves at the time.

Over in the real world, Campanella held out hopes that either the Spanish monarchy or the papacy would usher in a perfected civilization. This civilization would be resolutely Catholic, and it would tolerate no heresy whatsoever, which is to say that all would believe exactly as Campanella believed. That Campanella himself spent three decades in prison for heresy seems not to have dulled his enthusiasm for the project.[27]

As with Plato, the desire for a perfect social order came without any credible demonstration of the technology needed to support it. Indeed, it came well before anything like our modern scientific knowledge, which casts permanent doubt on astrology itself. Campanella's boldness—and the boldness of so many planners like him—should give us pause about the temptation to comprehensive planning. We want this planning without possessing its foundations; when the foundations are lacking, we imagine that we have them anyway.

It would be interesting, but it would take us far afield, to ask how Campanella came to believe that astrology was infallibly true. Certainly empirical evidence could not have supported this belief; even in his time, failed astrological predictions were well known. And, as we have seen, the church fathers, with Augustine at their head, scorned astrology, which smelled to them of superstition. It smells to us of dogmatism, because even a well-evidenced and generally reliable body of knowledge should remain open to doubt and revision, as skeptics from David Hume to Karl Popper have aptly noted. Solaria's astrology was not such an open system. If the Solarians *ever* miscalculated, even once, or if they ever abandoned weaker methods of astrology for more discerning ones, Campanella does not mention it. The technology that underpins the system appears outside the realm of dispute.

Also in the sciences of government, a degree of openness to doubt is a desirable quality, as are the capacities to conduct experiments and learn from mistakes. Paradoxically, without these qualities, it is unclear how one ever increases one's certainty in one's knowledge, much less how one arrives at the all-encompassing certainty of Solaria. These are matters somewhat outside our inquiry, though certainly related to it: When a purpose to government is articulated, how may it be tested and revised?

BOSSUET: THE GREAT CHAMPION OF ABSOLUTISM

Among those who believe that the purpose of government is to pattern society on something higher or greater than itself, there are many answers as to what this higher or greater thing might be. As we have seen, Plato had his numerology, and Campanella had astrology. But Jacques-Bénigne Bossuet turned to the Bible, and to a particular metaphysics that had been derived from it and from Platonic philosophy. One of the virtues of reliance on the Bible as a template for governance is that it makes utopianism in the style of Campanella more difficult—outside the Kingdom of God, original sin usually denies us the possibility of perfection. Yet the common interpretation of the Bible in Bossuet's day was nonetheless amenable to an expansive view of government, and nearly to an all-encompassing one.

Bossuet was the Bishop of Meaux, the court preacher to Louis XIV and tutor to Louis' eldest child, the Dauphin, or crown prince of France. American Christians often claim that the Bible supports a republican form of government with a significant degree of individual liberty, broadly free-market institutions and elements of representative democracy. Bossuet would have rejected each of these as completely unfounded. For him, the Bible recorded that absolute monarchy was the only form of government that God himself ever chose to establish; the Bible likewise supplied clear reasons for God's choice, and it showed why all other types of government were inferior.

Bossuet's case is much stronger than one might like to admit. He built it above all in two works: the *Discours sur l'histoire universelle* (*Discourse on Universal History*), published in 1681, and the *Politique tirée des propres paroles de l'Écriture sainte* (*Politics Drawn from the Very Words of Holy Scripture*), published posthumously in 1709. Let us consider them in turn.

The *Discourse* was intended to help teach the Dauphin history. It revealed much about Bossuet's theory of government, which he believed could be demonstrated with reference either to sacred or to profane history. Bossuet argued that rulers and states that followed absolutist principles were successful, while those that diverged were not. As a result, he held that the knowledge of history was useful for any ruler of "this great kingdom, which you [the Dauphin] are obliged to make happy."[28] Modern historians will find much to condemn in Bossuet's narration of specific historical events and in his judgments of rulers and epochs. Yet as a window into seventeenth-century absolutist political theory, the *Discourse* is enlightening indeed, and it shows that absolutists could at times be relatively modern in their thinking.[29]

Consider happiness: To the ancients, happiness might be the result of virtue, but happiness was not the purpose of government per se. And Augustine would hardly have been able to imagine a Christian ruler whose object was the earthly happiness of his people; bringing them to heaven was hard enough. Indeed, a significant Augustinian strain of thought existed in Bossuet's time, one that considered the advent of Constantine, the first Christian emperor, to have been a calamity: Constantine's conversion, while laudable like any other, had unfortunately married the Church to earthly things. Before Constantine, the primitive Church had been a relatively sure path to salvation; thanks to the persecutions, few would come to the Church *except* for good motives. But after Constantine, people would come to the Church for worldly reasons. To use Augustine's terms, the City of Man had invaded the City of God.[30]

In taking earthly happiness as a goal of government, and in proposing to attain it by natural methods, Bossuet appears surprisingly modern. Yet his prescriptions for state organization and state action were not. The *Discourse*'s treatment of ancient Rome is a good illustration here, both because it is representative of the work's inclinations as a whole and because we have already been following several other authors' judgments about Rome. Bossuet evaluated Rome according to French absolutist standards, and by them, he found it wanting.

As Bossuet would have it, Rome suffered whenever power was divided or shared in any way at all. Nearly all of Rome's difficulties could be traced, he thought, to the moments when it opted for something other than a strictly hierarchical state with a well-defined order of succession. These instances of divided or uncertain power were in his view a more serious enemy than Carthage, the Gauls, or Persia. Consider Bossuet's narration of an important event of early Roman history, the First Secession of the Plebes in 494 BCE:

> Rome, which had been so well defended from foreigners, now pondered whether she should perish by her own hand: jealousy had been awakened between the patricians and the people: the consular power, although already moderated by the law of P. Valerius, still seemed excessive to this people, too jealous of their own liberty. They retreated to the Mount Aventine [Mons Sacer]: councils of violence against them were of no use, and the people could not be brought back but by the peaceful remonstrances of Meninus Agrippa; but it was necessary to make accommodations to them, and to give the people the tribunes to defend them against the consuls.[31]

This passage is dead wrong about several key facts. The Roman countryside was by no means "well defended" at the time: The Volscians had just overrun much of it, and they were threatening to sack Rome itself—a state of affairs that one would never have guessed from relying on Bossuet. Bossuet simply and conveniently forgot the Volscians—until, narrating the events of some years later, he could describe them as instruments of an internal dissent: "Rome defeated all its enemies in the neighborhood," Bossuet wrote, "and seemed to have nothing to fear but itself. Coriolanus, the zealous patrician ... expelled by the popular party despite his services, pondered the ruin of his fatherland and brought the Volscians against it, reducing it to desperation, and he could only be appeased by his own mother."[32] In reality, the Volscians had been there all along.

Bossuet likewise condemned the institution of the tribunes, which Machiavelli had praised. The latter considered the tribunes a key constitutional stabilizing measure: By giving the common people a limited voice in government, they allowed a republican kind of *virtù* to flourish, one in which all social orders had a share. Republics that managed this difficult feat became strong, Machiavelli thought, and his prime example was Rome itself.

Bossuet disagreed and indeed condemned the tribunes, holding that they "increased divisions in the city; and Rome, formed under kings, lacked the necessary laws for the good constitution of a republic." What he thought these provisions might have been, he never revealed.[33]

Without a monarchy, Rome was doomed to constant dissension, particularly given the ambitions of its citizens and the crass appeal to popular vanity:

> But the more the face of the republic appeared beautiful from without, thanks to its conquests, the more was it disfigured by the unbridled ambition of its citizens, and by its internecine wars. The most illustrious Romans became the most pernicious to the public good. The two Gracchi, by flattering the people, began the divisions, which did not end until the end of the Republic itself ...
>
> Rome was torn apart ... by the furors of Marius and Sulla. Sulla ... did all too much against his fatherland, in that his tyrannical dictatorship reduced it to slavery. He could indeed voluntarily step down from sovereign power; but he could not prevent the effects of his bad example. Everyone wanted to dominate.[34]

Everyone wanted to be the sole master, and without a rule for succession, almost everyone could aspire to it. When everyone aspired, everyone plotted against the republic, rather than working harmoniously. The notion that a single, all-powerful executive produces stability is a common claim in Renaissance and early modern political theory, and we even have even heard similar claims in recent American politics, from those who bemoan our disunity and promote national greatness under a strong leader as the remedy. In this, Bossuet was typical of his time, and he might not have felt wholly out of place in our own.

As Bossuet saw it, the wars and conspiracies continued under the Empire as well, and for the exact same reasons: even then, no one knew their place in Roman society, and all continued to harbor the same dangerous ambition. There were good emperors, and under them the Empire prospered. But there were also bad emperors, and worse, there were times of interregnum, when no authority predominated.

The subtext here was that an absolute monarch with a clear rule of succession could put an end to discord. Then each person would be born into a place, and all would know what that place was, and none—it was hoped—would rebel against the king or the order he represented. "Is there any thing that the lust for power will not do in our hearts?" Bossuet asked.[35] Only absolutism could keep our lust in check.

For Bossuet, all of this was but a subset of the Great Chain of Being: God the Father was a monarch, and lesser rulers, like the earthly kings, acted by His grace and according to His purposes. Absolutism checked the pride of all people, save possibly one, and that one was accountable before God.

Bossuet thought that Rome itself had received a foretaste of all this at the time of Augustus. "Victorious at sea and on land, [Augustus] shut the temple of Janus. All the universe dwelt in peace under his power, and Jesus Christ came into the world."[36] Christ's arrival at such a time was no coincidence; as the Prince of Peace, Christ had only deigned to appear as a man under the peace of a proper absolutism. It was a golden age in other ways as well: "All the arts flourished in [Augustus'] time, and Latin poetry was carried to its ultimate perfection by Virgil and Horace," of whom Augustus had been patron. The principle of hereditary succession was likewise established, added Bossuet, who even put in a good word for Augustus' justly reviled stepson, Tiberius: "Rome had to suffer much under the cruel reign of Tiberius," he wrote, "but the rest of the empire was tranquil enough."[37]

A further illustration of this attitude can be found in Bossuet's discussion of the Hellenistic period in Judaism. He viewed it as a fortunate time,

one whose strengths derived from the unity of the people under a strong
central state with a relatively orderly succession:

> The Jewish religion was known among the gentiles. The temple of Jerusalem
> was enriched by the gifts of kings and peoples. The Jews lived in peace and
> in liberty under the power of the kings of Syria, and they had hardly known
> such a tranquility under their own kings. It would seem set to last forever,
> if the Jews themselves had not disturbed it through their dissensions ...
> Certain of the most powerful betrayed their people to flatter the kings; they
> wished to make a name for themselves in the manner of the Greeks, and they
> preferred this vain pomp to the solid glory that would have accrued to them
> among their fellow citizens [*citoyens*] through the observance of the laws
> of their ancestors ... As is typical, the jealousies and divisions of individuals
> took little time in causing great harms to all the people.[38]

What should Christians do when faced with an unjust ruler? Bossuet's
answer was simple: They should submit. "[N]ever did [the early] Christians
cease to respect the image of God in the princes who persecuted the truth
... [T]hey prohibited among themselves not simply seditious actions,
but even murmurings. The finger of God was in this work." He claimed
that submission was the direct consequence of the injunction found in
Matthew 22:21 to "Render therefore unto Caesar the things which are
Caesar's; and unto God the things that are God's."[39]

Americans often use these words to justify the separation of church and
state. Here they justify obedience to an evil government. This deployment
will strike today's readers as strange and wrong. Yet it was not so strange in
other times and places. On the contrary, theorists of absolutism commonly
used Matthew 22:21 in precisely this way: What belonged to God was the
soul and all matters pertaining to divine worship. Caesar, meanwhile, got
absolutely everything else.[40] Far from being a roadmap to religious liberty,
this reading of "render unto Caesar" extinguished every form of liberty
in the secular sphere, while leaving only a single religion to choose from.

All that Bossuet thought he had discerned in secular history, he likewise
saw in the words of scripture. As his nephew, the abbé Bossuet, put it in
the introduction to *Politics Drawn from the Very Words of Holy Scripture*:
"The authority on which everything herein is based could not be more
sacred, nor more incontestable: it is that of the Holy Scripture, and of He
who speaks throughout it, that is, the Sovereign Master of Kings."[41]

The person of a king is sacred, wrote Bossuet, drawing on the ritual of
anointment narrated in 2 Kings 9. To attack a king is not merely an act of

assault, or of treason; it is a sacrilege.[42] Kings represent the divine majesty, sent down by God's providence to realize his designs.[43] Kings are therefore "christs," in the original sense of the word; they are God's anointed, even if they are pagan: For God, speaking through Isaiah, had called Cyrus, a pagan king, to be his anointed.[44] All kings, even evil ones, carried this status, which placed them high indeed in the Great Chain of Being.

It only followed that Bossuet denied the existence of a natural equality among men. While it was true for him that all men were brothers, kings nonetheless held a distinct status among them.[45] In this denial of natural equality, he was quite unlike, for example, Thomas Hobbes, and much closer to the consensus view among the learned absolutists of his day.

States were also part of the natural order of things, and among them, absolute monarchy was the highest and best type of state. We could know this because God gave his chosen people a monarchy, and because He likewise reigns as an absolute monarch in Heaven. And what God established for his chosen people in the Old Testament, He reaffirmed in the new. Bossuet reviewed many of the standard New Testament justifications for the divine right of kings, including Romans 13[46]:

> Let every soul be subject unto the higher powers. For there is no power but of God: the powers that be are ordained of God.
> Whosoever therefore resisteth the power, resisteth the ordinance of God: and they that resist shall receive to themselves damnation ...

Colossians 3:22–24:

> Servants, obey in all things your masters according to the flesh; not with eyeservice, as menpleasers; but in singleness of heart, fearing God;
> And whatsoever ye do, do it heartily, as to the Lord, and not unto men;
> Knowing that of the Lord ye shall receive the reward of the inheritance: for ye serve the Lord Christ.

And 1 Peter 2:13–18:

> Submit yourselves to every ordinance of man for the Lord's sake: whether it be to the king, as supreme;
> Or unto governors, as unto them that are sent by him for the punishment of evildoers, and for the praise of them that do well ...
> Servants, be subject to your masters with all fear; not only to the good and gentle, but also to the froward.
> For this is thankworthy, if a man for conscience toward God endure grief, suffering wrongfully.

Those who rebelled were not merely public enemies. They were also the enemies of God.[47] The reasoning went something like this: Good rulers were to be obeyed, because a Christian was obedient to the good. Bad rulers were to be obeyed, not for love of evil, but because Christians were to accept lovingly the chastisements of God, and among these chastisements were bad rulers. Christians were to be *thankful* for the grief that they endured by way of conscience; that, too, was of God.

Where exactly did this leave the individual? Bossuet answered: "One must be a good citizen and sacrifice to [the sovereign] in time of need everything that one has, including one's own life ... All the love that one has for oneself, for one's family and friends, is reunited in the love one has for one's fatherland [*patrie*], in which our happiness and that of our families and friends is reunited."[48] The state's interests already were one's own—one simply had to recognize them as such, and all would be well. "He who serves the public serves each individual," Bossuet wrote.[49]

In the *Politique*, Bossuet developed even more clearly the idea that we have shown in the *Histoire Universelle*: "Only the authority of government can check the passions and the violence that has become natural among humans ... It is only by the authority of government that union is established among men," he wrote here.[50] Violence was natural among humans owing to original sin. Government, like all good things, was a defensive bulwark against sin. The passion for domination was to be subjugated, and only the ruler was to have dominion.

Most likely thinking of Joshua 4, Bossuet wrote:

> [Israel] was forty thousand men, and all this multitude was as one. That is the meaning of the unity of a people, in how each, renouncing his own will, transports and reunites it with the will of the prince and the magistrate. Otherwise there is no union; the people would have been vagabonds, like a scattered herd.[51]

Now, those who reject absolute monarchy are by no means compelled to adopt Bossuet's interpretations of scripture. Even in his day, a diverse literature had developed in which the Bible justified rebellion against wicked governments. Discussing that literature in detail is beyond the scope of this book.[52] For our purposes, it is enough to say that Bossuet disagreed. Yet what about those points of scripture in which God's people do engage in holy revolts? What did Bossuet make of them?

The episode of the Maccabees did nothing to justify any other rebellion, Bossuet claimed:

> It is not permitted to doubt that the Maccabees' revolt was just, for God himself approved of it. But if we consider the factual circumstances, we will find that they do not warrant the revolts that religious motives have occasioned among us.[53]

God had definitionally good reasons to approve of the revolt, and these Bossuet tried to explain. The Seleucid king Antiochus IV had not only commanded the Jews to perform idolatry, but he had punished refusal with death. Frustrated with the Jews' intransigence, Antiochus sacked Jerusalem, massacred Jews by the thousands, and sold many of the rest into slavery. His wickedness was impossible to deny.

But was he not still a ruler? And did he not therefore deserve obedience? To solve this all but impossible dilemma, Bossuet declared that Antiochus had "abdicated." "Could one more overtly renounce one's having [the Jews] as subjects?" he asked. This claim must have made Bossuet squirm, because by the usual meaning of word, Antiochus had done nothing of the kind. He had not abdicated. He had simply been a tyrant. This odd usage brought Bossuet, absolutist though he was, perilously close to the logic of the regicides in the English Civil War. Yet Antiochus was hardly the only bloodthirsty tyrant ever to have ruled, and not even the only one to have oppressed the chosen people. What about pharaoh, or Nero, or Diocletian? Were they not analogous? Wouldn't revolt have been justified against them as well?

These questions call into doubt Bossuet's entire theory of government. And many more exist: What, for example, of King David, who rebelled against Saul? David had been unique in all of history, Bossuet answered, because David was God's anointed—and yet, we might ask, if God's anointed can be a rebel, then isn't rebellion at least sometimes licit?[54] And hadn't God chosen Saul as well, through a direct revelation to the prophet Samuel? If one could rebel even against a king who had been chosen by God, then how can we distinguish good revolts from bad ones? Easy answers to these questions probably do not exist.

Bossuet's frequent appeals to divine authority have no doubt prevented him from receiving the sort of attention that political philosophers have given to Hobbes. Yet Bossuet's account of absolutist ideology was in many ways truer, I think, to seventeenth-century practice and conventional wisdom than Hobbes', and perhaps Bossuet should therefore not be so neglected.[55] Bossuet's absolutism was founded squarely in Christianity, and not on the newfangled Cartesian psychology found in Hobbes; the latter was likely as alienating to Hobbes' own readers as it can be to us. (For our own treatment of Hobbes, we will simply ignore the psychological aspects of his thought, as many political theorists have profitably done. The rest stands well enough on its own.)

Bossuet's strength for us is that he did not theorize a thing he had never seen; he actually lived in the court of a well-established absolutist regime, and he straightforwardly described the principles that he saw professed, and enacted, all around him. None other than Louis XIV had tasked him with transmitting absolutist values to the rising generation, and by all evidence Bossuet's patron was pleased with his work.

Bossuet had little use for Robert Filmer's comparatively weak claim that kingly authority had been inherited—somehow—from Adam's patriarchal authority. Although to Bossuet patriarchal authority was the "first" form of kingship, others were also legitimate. The relationship was less one of inheritance and more one of aiming at a shared ideal. As a result, kings who were popularly acclaimed, or who came by their titles through conquest, were no less legitimate than patriarchal kings. They also posed no great theoretical problems to him. There was no need to worry about which particular man in all of the country, or all the world, was the proper heir to Adam, an embarrassing difficulty for which John Locke had ridiculed Robert Filmer in Chaps. X–XI of the *First Treatise*.[56]

Kings came from a variety of sources, held Bossuet, and many were indeed kings by nothing more than acclamation. It was not the case that the multitude had somehow identified from among themselves the one true heir of Adam. It was simply that each individual knew that it was better to live under a king, and to establish an orderly rule of succession thereafter.

Yet unlike Hobbes, Bossuet was in no sense a social contractarian. No founding agreement had ever taken place *among equals*, whether in historical time or metaphorically, and no state of nature had ever existed to be pondered upon or exited. For Bossuet, the acclamation of a king was simply the recognition by a multitude that God and nature favored monarchy, as well as the recognition that one individual had been chosen by an institution of divine providence (and *not* by an agreement), to rule. The people did not agree to a king; they merely recognized one.[57]

In Hobbes' system, all men were born equal, and they gave up their natural equality to form a civil society. In the process, they surrendered much liberty; in return, they received the (artificial) security of living under a sovereign. By contrast, Bossuet had no use for the natural equality of mankind. It was clear enough to him that men had vastly different abilities, proclivities, virtues, and vices. His system had no need whatsoever to deny it. Meanwhile, monarchy was by no means artificial; on the contrary, he held it "the most natural" form of government.[58] As a result, he likewise found monarchy "the most durable ... and the strongest."[59] Moreover:

> It is also [the form] the most opposed to the division that is the most essential evil of states, and the most certain cause of their ruin ... When we form states, we seek to unite ourselves together, and we are never more united than when we are under a single chief. Never are we stronger, either, because all is done in harmony [*parce que tout va en concours*].[60]

A war of all against all was *not* for Bossuet the state of nature—a war of all against all was merely *evil,* and evil was for him a part of the human condition ever since the Fall. Yet redemption was also a part of the human condition, and the way to redemption from this particular evil was already well known in the time of Solomon. It consisted of adopting an absolute monarchy. The progress from an ungoverned to a governed state was not a progress from nature into artifice. It was a progress from wickedness into conformity with God's law.

Absolutism thus proposed to solve the problem of coordinating human ambitions—which was a facet of the problem of original sin. Because absolutism allegedly solved this problem, we should choose absolutism in our own societies. And, not coincidentally, God had ordained it as well.

Bossuet's influence on the anglophone world appears to have been small, but it remains a topic for further study, which might reveal it to have been greater than it now seems. In particular, Bossuet's argument for hereditary monarchy bears a strong resemblance to one found in Edward Gibbon's *Decline and Fall of the Roman Empire*, book I, Chap. 7:

> Of the various forms of government which have prevailed in the world, an hereditary monarchy seems to present the fairest scope for ridicule .... but our more serious thoughts will respect a useful prejudice, that establishes a rule of succession, independent of the passions of mankind; and we shall

cheerfully acquiesce in any expedient which deprives the multitude of the dangerous, and indeed the ideal, power of giving themselves a master …
   The superior prerogative of birth, when it has obtained the sanction of time and popular opinion, is the plainest and least invidious of all distinctions among mankind. The acknowledged right extinguishes the hopes of faction, and the conscious security disarms the cruelty of the monarch. To the firm establishment of this idea we owe the peaceful succession and mild administration of European monarchies.[61]

Both Bossuet and Gibbon seem to have found hereditary monarchy a fitting antidote for Rome's particular instability.
   In the language of present-day political theory, Bossuet chose absolutism based on two claims, one of value and one of fact. First, he has made the value claim that something like the Rawlsean *stability criterion* is the most important one for the evaluation of various constitutional types. As John Rawls put it:

Since a well-ordered society endures over time, its conception of justice is presumably stable: that is, when institutions are just … those taking part in these arrangements acquire the corresponding sense of justice and desire to do their part in maintaining them. One conception of justice is more stable than another if the sense of justice that it tends to generate is stronger and more likely to override disruptive inclinations and if the institutions it allows foster weaker impulses and temptations to act unjustly.[62]

A relatively stable theory of justice will foster only relatively weaker impulses and temptations toward injustice. Participants in a society constructed according to a relatively stable theory of justice will be more likely to put aside whatever disruptive inclinations they might personally harbor. Second, Bossuet made the factual claim that absolute monarchy best satisfies the criterion. Among all governmental types, absolutism best instills loyalty to itself.
   These are strange moves, to be sure, but they are moves within a game whose rules we recognize: Were we to settle on a similar ranking of priorities, we would be forced to conduct a similar empirical search for the stablest government. We might reach different results, of course, but these we might bring to one of Bossuet's followers with the expectation of a reckoning. We would be playing the same game that he did.
   As a result, I do wonder whether in setting up Hobbes and Filmer as the great exponents of absolute monarchy, the anglophone world has not

built up a pair of strawmen, and avoided the most challenging exponent of the ideology. Hobbes was simply too eccentric to make a good contemporary representative of absolutism, and, as we shall see, he had some important liberalizing tendencies that were absent in Bossuet. Filmer, meanwhile, might have made a much stronger case had he jettisoned the clearly unworkable framework of inherited patriarchal authority. With no qualms at all, Bossuet did just that. His reasoning is frequently secular in its operation, but it also aims to explain the apparent divine preference for absolutism. God turns out to have had good reasons indeed, and we are fully capable of understanding them, says Bossuet. Divine right is perhaps more than divine fiat.

Yet Hobbes was an important figure in the story we have to tell, albeit for reasons other than his defense of absolutism. It is to his particular importance that we now turn.

NOTES

1. Skinner, "The State" *in* Terence Ball, James Farr, and Russell L. Hanson, *Political Innovation and Conceptual Change*, pp. 90–131. Skinner notes that the etymological ancestors of the term "state" referred originally not to an institution, but to the *status* of individual rulers. By Machiavelli's time, the modern sense was common, helped along by the revival of Roman law and the concept found within it that the republic had a *status* as well.
2. Berlin, "The Question of Machiavelli," *New York Review of Books*, November 4, 1971, reprinted in Nicolò Machiavelli, *The Prince*, Norton Critical Edition, 2nd Ed., Trans. Robert M. Adams, 1992, p. 214.
3. Objections to this line of reasoning are many and easy to name, including the notion that the good might entail some flexibility. Or that the admixture of falsehood ensures the impermanence (and thus the badness) of the state. And even if we grant that everything good is permanent, it does not follow that everything permanent is good.
4. Machiavelli, *Discourses*, p. 139.
5. Ibid., p. 140.
6. Ibid., p. 139.
7. Ibid., p. 143.
8. Ibid., p. 145.

9. Machiavelli, *The Prince*, p. 42.
10. Machiavelli, *Discourses*, p. 517.
11. Hume, "Of Civil Liberty," in *Essays*, p. 87.
12. Cassirer, *The Myth of the State*, p. 140.
13. Popper, *The Open Society and Its Enemies*.
14. Before Campanella, of course, there was Thomas More (1478–1535), whose *Utopia* (1516) is probably better known. More's book is more a puzzle than a polemic, however; it is full of ambiguity and complexity, and it lacks the earnestness of Campanella's utopia. Among those that seem to incline more in Campanella's direction are Francis Bacon's *New Atlantis* (1627), James Harrington's *Oceana* (1656), Louis-Sébastien Mercier's *The Year 2440* (1770), Edward Bellamy's *Looking Backward* (1888), and numerous more recent works, particularly in the genre of science fiction. Political theory seldom looks closely at most such works, but given their influence on the public mind, perhaps it should. For the wide influence of the now nearly forgotten Mercier, see Robert Darnton, *The Forbidden Best-Sellers of Pre-Revolutionary France*.
15. Tommaso Campanella, *The City of the Sun*, p. 87.
16. This site helps with some of the calculations: http://www.configurationhunter.com/astrology-tools/planet_positions.
17. Daniel J. Donno, in Campanella, pp. 133–34, n 36.
18. Campanella, p. 37.
19. Ibid., pp. 55–57.
20. Ibid., p. 27.
21. Ibid., p. 41.
22. Ibid., pp. 33–37.
23. Ibid., p. 81.
24. Ibid., p. 83.
25. The most important work on this mode of thought, usually styled civic humanism, is Pocock, *The Machiavellian Moment*.
26. Headley, "On the Rearming of Heaven."
27. Donno, "Introduction," in Campanella, pp. 2–15. That Campanella defended Galileo from the charge of heresy is immaterial; Campanella would have punished plenty of other accused heretics—just not this one.

28. Bossuet, *Discours sur l'histoire universelle*. http://www.samizdat. qc.ca/cosmos/sc_soc/histoire/hist_med/hist_universel.pdf, p. 8. All translations from Bossuet by the author.
29. Hulliung, "Patriarchalism."
30. Kuznicki, *Reasonable Souls*. See also Dale K. Van Kley, "Civic Humanism."
31. Bossuet, *Discours*, p. 31.
32. Ibid.
33. Bossuet, *Politique*, p. 67.
34. Bossuet, *Discours*. p. 47.
35. Ibid., p. 56.
36. Ibid., p. 49.
37. Ibid.
38. Ibid., p. 115.
39. Ibid., p. 166. Bossuet here cites Matthew 22:21.
40. Chief among them is Filmer's *Patriarcha*, in which the passage was discussed on two occasions. Both defended absolutism in all matters not touching on religious practice. They were offered not to limit royal power, but to establish political obligation in everything else. Filmer, *Patriarcha*. http://oll.libertyfund.org/titles/221.
41. "Epistre à monseigneur le dauphin," by the abbé Jacques-Bénigne Bossuet (bishop of Troyes), *in* Jacques-Bénigne Bossuet, *La Politique tirée des propres paroles de l'Écriture sainte*. Paris: Pierre Cox, 1709. 10th unnumbered page.
42. Bossuet *Politique*, p. 82.
43. Ibid., p. 83.
44. Ibid., p. 83.
45. Bossuet also believed that succession in the male line was the best rule to follow. *Politique*, p. 71.
46. Bossuet, *Politique*, pp. 84–5.
47. Ibid., p. 76.
48. Ibid., p. 42. Hegel might have said that the love for self, family and friends was *sublated* into the love for the fatherland—the idea seems virtually the same—but Bossuet lacked the gift of obscurantism by which preposterous ideas are made to seem profound.
49. Ibid., p. 43.
50. Bossuet, *Politique*, Article III Prop. 2–3.
51. Bossuet, *Politique*, p. 23. The marginal citations are to 1 Kings 11 and 7 and to 1 Esdras 2, neither of which seem apposite.

52. Skinner, *The Foundations of Modern Political Thought*, particularly vol. 2, part III.
53. Bossuet, *Politique*, p. 279.
54. Bossuet, *Politique*, p. 275.
55. To name just one example, Bossuet's absolutism resembles that of the immature John Locke. See Peter Laslett's introduction to *John Locke: Two Treatises of Government*, pp. 19–20. Laslett's observation is also apt: "[Hobbes'] influence on thought about politics has been enormous, but his purchase over what men do politically has been negligible." In ibid., pp. 90–91.
56. Locke, *Two Treatises of Government*, p. 216.
57. Bossuet, *Politique*, p. 64.
58. Ibid., p. 67.
59. Ibid., pp. 69–70.
60. Ibid.
61. Gibbon, *Decline and Fall*, http://oll.libertyfund.org/titles/1365#Gibbon_0214-01_346.
62. Rawls, *A Theory of Justice*, p. 398.

CHAPTER 6

The Social Contractarians: Can an Agreement Specify What Government Is For?

To review, one common purpose that has been offered for government is that it exists (or should exist) to align human society, otherwise messy and chaotic, with a transcendent or universal order. This type of answer most clearly encompasses Plato, Campanella, and Bossuet, with the caveat that each had a different transcendent order in mind. Many other authors could be added to this list, including some whom we will meet later on.

The faults of this approach are apparent. It is unclear whether we down here on earth can access a transcendent order at all, and if we can, it is a further question whether our access to it can provide an appropriate model for human societies. What is clear is that many people feel that such an order *should* exist, complete with political implications, and perhaps that some have already personally accessed it. And yet when the prophets return from the mountaintops, they so often disagree on what they have seen there.

A second proposed purpose for government is that it is to serve as a vehicle for the exercise of a set of virtues, usually the martial and self-sacrificing ones. The state becomes a theater for these virtues' presentation and scrutiny, and their good exercise is nearly an end in itself. I do say nearly, because on this view, the state is apt to become the highest end instead. The ancient historians almost uniformly shared this approach, and Machiavelli certainly did. Like the transcendent order, what I have termed the "theater of virtue" will never entirely disappear; it will come up again on several further occasions in this book and, indeed, it can still be found in present-day political thinking.

© The Author(s) 2017
J. Kuznicki, *Technology and the End of Authority*,
DOI 10.1007/978-3-319-48692-5_6

It will remain, however, a poisonous idea: As we shall discuss, the presence of martial virtues does not necessarily make for a stable or orderly government. Such virtues may be ardently desired ex ante, but ex post, one finds that their demonstration often leads to an endless contest over which leader or faction represents the truest and most virtuous actor in virtue's theater. Such was the case with the French Revolution.

The third major category of answer to our focal question is that government exists to resolve conflicts in our values, laws, and customs, with the eventual goal of bringing earthly laws into harmony with one another and also into accord with justice—not in one fell swoop, but rather in the fullness of time. If this is an alignment with a transcendent order—which one might argue—it will always be an imperfect one. Yet again, we are not done with this answer; we will encounter it on several further occasions.

And, finally, government may have *no legitimate purpose at all.* It may simply be an affliction, one whose only purpose, if we may call it that, is to serve as a chastisement from God. Examples of this category come from both pagan and Christian sources, and, like the others, we will meet it again.

We now turn to the social contract. Intellectual historians have advanced many definitions of social contractarianism over the years, but the one I favor is simple: Social contract theories are those that hold that government is (at least in some important cases) man-made, conventional, or artificial, and that it is produced by some form of agreement, or by something sufficiently like an agreement to make for a useful analogy.

Social contract theory need not deny or exclude a role for divine providence. The Christian and Jewish traditions alike have always stressed that God instituted governments by *covenant,* and thus by an agreement. Government by conquest, though, is generally either downplayed, delegitimized, or else retrospectively likened to an agreement. Hobbes, for instance, argued that agreement on pain of death was nonetheless a form of agreement—a solution that few will likely find satisfying. His move seems to open up the prospect of varying degrees of legitimacy within the contractual framework, a possibility he nonetheless denied.

Whatever the case may be, social contract theorists raised many new questions that are closely related to our focal question, albeit not identical to it. For example: As a matter of historical fact, how were the purposes of actual governments agreed upon? If these historical processes were blameworthy, then how *should* the purposes of government be agreed upon? Are rulers ever accountable to authorities other than God? If so, to whom, and how, and for what? What happens when the relevant parties are dissatisfied with the rulers' performance? What about those governments that clearly were not established by agreement—can they have legitimate purposes? These questions all arise from the recasting of political theory as the analysis of a contract, compact, or agreement.

Describing government as an agreement among people was in no sense an idea that originated with Hobbes and Locke, however. The idea can be traced to many different sources prior to our prominent seventeenth-century authors. Reviewing the origins of social contract theory, Harro Höpfl and Martyn P. Thompson cite "Old Testament history; medieval biblical exegesis; medieval references to contracts and feudal understandings of social and political relations; the teachings of the Church Fathers; the ideas of the Conciliar Movement; Lutheran and Calvinist ideas and practices; texts of the civil law ... the gradual breakdown of the extended family; the rise of capitalism; the rise of individualism and rationalism; and the impact of the scientific revolution of the seventeenth century."[1]

As an early modern historian might say: the usual suspects. A few examples are helpful in illustrating the broad origins of social contract theory before the seventeenth-century greats. First, two Jesuit scholastic philosophers, Luis de Molina (1535–1600) and Francisco Suárez (1548–1617), both justified the existence of positive, or man-made law, by reference to the conditions that would arise without it. As Quentin Skinner summarized their argument, "had we continued to live in our natural and pre-political communities ... we should soon have found our lives gravely impaired by increasing injustice and uncertainty."[2] Whether God had ordained any particular king turned out to be immaterial. *Some* king was better than *no* king, and reasonable people would agree to institute one, in light of the alternative. Spiritual matters had little to do with it, and thus the transition to a governed society was increasingly understood as a movement *out of* a natural state of being, and *into* an artificial one, a state made by man.[3]

The idea that society, including the state, could be the product of conventions, is one that can be found as far back as the Sophists. But apart from the Epicureans, ancient conventionalists tended to see government as the product of *harmful* conventions, which divided people into slave and free, or Greek and Barbarian. Nature did not distinguish in these ways, claimed Antiphon and Diogenes. One of the hallmarks of early modern social contract theory was to revive the Epicurean idea that there might be *helpful* social conventions—artificial social practices, including perhaps government, that were nonetheless beneficial.

The social contract was also not entirely theoretical. In the medieval era, many forms of government were literally founded in compacts, which had been agreed upon at a particular place and time, by real people who might have chosen otherwise.[4] Late medieval town governments commonly acted on the authority of a charter, one agreed to both by the inhabitants and the agents of government. These charters often recognized extensive liberties for the *bourgeois*, that is, for the town dwellers. Towns had independent court systems, commercial regulations that aimed at fair and transparent business practices, and even—one might say—liberal immigration policies: It was common that one could claim bourgeois status, along with its many legal protections, simply by being a resident for a year and a day, during which time, presumably, one would not have made oneself obnoxious.

Along with scholastic philosophy, charter oaths may lie at the origin of modern social contract theory. But in the case of the English social contract tradition, the roots are buried deep: Hobbes and Locke were Protestants; Molina and Suárez were not only Catholics but also Jesuits. And the Jesuits were particularly hated in seventeenth-century England for their attempts, suspected and sometimes genuine, to overthrow the English government. Hobbes also detested Suárez on a variety of other grounds.[5] As to charter oaths, these were far removed from the purposes that Hobbes and Locke had set out for themselves. Both writers aimed at an account of political power that would work for nation-states, not for subdivisions thereof.

Moreover, nation-sized social contract theory was already alive and well much closer to home: It could be found in the writings of the Levellers, who were Protestant proto-libertarian activists in 1640s. It could also be found coming from the poet—and political theorist—John Milton, whose tract *The Tenure of Kings and Magistrates* (1649) used social contract theory to justify the execution of King Charles I:

> Trajan the worthy Emperor, to one whom he made General of his Prætorian Forces [said] "Take this drawn sword ... to use for me, if I reigne well, if not, to use against me." ...And not Trajan onely, but Theodosius the yonger, a Christian Emperor and one of the best, causd it to be enacted as a rule undenyable and fit to be acknowledg'd by all Kings and Emperors, that a Prince is bound to the Laws; that on the autority of Law the autority

of a Prince depends, and to the Laws ought submitt. Which Edict of his remains yet in the Code of Justinian l. I. tit. 24. as a sacred constitution to all the succeeding Emperors. How then can any King in Europe maintain and write himself accountable to none but God, when Emperors in thir own imperial Statutes have writt'n and decreed themselves accountable to Law. And indeed where such account is not fear'd, he that bids a man reigne over him above Law, may bid as well a savage Beast.[6]

Rather than seeking any definitive origin, it may be easier simply to say that social contract theory was in the air at the time, and that western culture, and Britain particularly, had been awaiting thinkers who could more clearly explore some of its surprising implications.

HOBBES: ABSOLUTE POWER AND ITS LIMITS

In the one instance that Thomas Hobbes discussed town charters, he seemed at pains to downplay their importance and stressed that they were merely grants of privilege from the sovereign, who could always revoke them at his pleasure.[7]

Elsewhere he mentioned overpowerful towns as an evil because they tended to undermine the absolute authority of the sovereign.[8]

But why settle on *absolute* authority? This question gets us to the heart of Hobbes' reasoning about government, which we will now take up. The purpose of establishing a sovereign, Hobbes claimed, was to check individual violence. Leaving it unchecked, as it would be in the state of nature, would lead to a much more disagreeable condition for everyone. The passage in which he describes this condition is justly famous and often quoted:

> the nature of war, consisteth not in actual fighting; but in the known disposition thereto, during all the time there is no assurance to the contrary...
>
> In such condition, there is no place for industry; because the fruit thereof is uncertain: and consequently no culture of the earth; no navigation, nor use of the commodities that may be imported by sea; no commodious building; no instruments of moving, and removing, such things as require much force; no knowledge of the face of the earth; no account of time; no arts; no letters; no society; and which is worst of all, continual fear, and danger of violent death; and the life of man, solitary, poor, nasty, brutish, and short.[9]

In this, Hobbes' system resembled the absolutism Bossuet would later articulate, which we have already examined. Government exists in part to put aside certain possibilities, so that others may come to the fore, including cultivation, industry, arts, letters, and society as we usually understand it. Hobbes epitomized the covenant that established a government in the following terms:

> I authorise and give up my right of governing myself, to this man, or to this assembly of men, on this condition, that thou give up thy right to him, and authorize all his actions in like manner.[10]

It remains to be shown, one might object, that *absolute* or *irrevocable* power is the right tool for the job. An absolute or irrevocable form of government may indeed be neither necessary nor even sufficient to provide the security that Hobbes wanted. It is trivial nowadays to observe many countries living more or less peaceably under governments that are neither absolute nor irrevocable, that tolerate much in the way of dissent, that contain various competing power structures, and that even countenance— as with the U.S. Declaration of Independence—the right of revolution. Indeed, absolute power may deliver less security; one seldom observes so much fear of death, so much instability, so much war, and so little in the way of industry or culture, as under a totalitarian government. Such governments have seized (irrevocably, at least they say) all of the powers that Hobbes would have given to a sovereign.

Hobbes' theory nonetheless represents a step in the direction of liberalism. That's because *Leviathan* ultimately argued that states are to have specific, articulated, and therefore—in principle at least—*limited* purposes. As a direct consequence, some things can be understood as clearly off-limits to the sovereign. Hobbes even called out some of them as improper. Whenever such impermissible purposes are advanced, individuals need not obey their sovereign. Concerning the agreement that established the sovereign, Hobbes wrote,

> No man is bound by the words themselves, either to kill himself, or any other man; and consequently, that the obligation a man may sometimes have, upon the command of the sovereign to execute any dangerous, or dishonourable office, dependeth not on the words of our submission; but on the intention, which is to be understood by the end thereof. When there-

fore our refusal to obey, frustrates the end for which the sovereignty was ordained; then there is no liberty to refuse: otherwise there is.[11]

But who would judge whether a killing was proper, and on what terms? Again, the central question of our own inquiry—*What is government for?*—is crucial to determining when the right to disobey is allowed. As the political philosopher Leo Strauss wrote of Hobbes' system:

> The state can justifiably demand from the individual only *conditional* obedience, namely an obedience that does not stand in contradiction to the salvation or preservation of the life of this individual; for the securing of life is the ultimate basis of the state ... The right to the securing of life pure and simple ... has fully the character of an inalienable human right, that is, of an individual's *claim* that takes precedence over the state and determines its purpose and limits.[12]

All individuals sought to preserve their own lives; government was a means to this end, and it properly restrained our natural liberty whenever our natural liberty threatened the lives of others. But when a government directly attacked us, it had ceased to keep its end of the bargain. It might likewise be irrational to keep with the sovereign when he commanded us to undertake the dangerous act of trying to kill someone else.

And so, for Hobbes, a man whose sovereign commanded him to kill was under no necessary moral or political obligation to comply with the sovereign's command.[13] The sovereign may order it, and he may even order it on pain of death, but the subject was (perhaps surprisingly) *morally free to resist*. Even a man duly condemned to death could not be called unjust for resisting his sentence, for justice itself was rooted in the preservation of life.[14]

Hobbes appears to have given this paradoxical conclusion considerable thought, and he seems more or less comfortable with the consequences it entailed, even when these consequences took him far away from the conventional wisdom. Elsewhere he wrote, "[E]very man ought to endeavour Peace as farre as he has hope of obtaining it; and when he cannot obtain it ... he may seek, and use, all helps, and advantages of Warre."[15] Even soldiers who deserted the battlefield were to be condemned not for injustice, Hobbes claimed, but only perhaps for cowardice. Protecting their lives was the whole point of government, meaning that even on the battlefield, resistance to government was justified by the very impetus that had called government into being in the first place.[16]

What other rights did subjects retain? As Quentin Skinner puts it, the list was "a remarkably extensive one."[17] It included not only the above mentioned rights of resistance but also the right to resist assaults by non-state actors, the right to refuse to accuse oneself, the right to "the use of food, ayre, [and] medicine" in spite of a sovereign's command to the contrary, and the right to refuse "any dangerous, or dishonourable Office."[18] Depending on one's definitions of "danger" and "dishonor," these last exceptions may overwhelm the rule.[19]

A further key to Hobbes' system was that a power legitimately granted, even over legitimate objects, would not always be exercised: When the laws were silent, the subjects would have liberty. That liberty was to be found in whatever the sovereign had "praetermitted"—that is, in whatever he and his laws ignored. No commonwealth could possibly legislate for everything, and none had ever attempted it. Among areas that might be praetermitted, Hobbes suggested "to buy, and sell, and otherwise contract with one another; to choose their own abode, their own diet, their own trade of life, and institute their children as they themselves think fit; and the like."[20] None of these can be called trivial liberties.

Finally, tucked away in a chapter at the tail end of the book, under the unpromising heading "of the benefit that proceedeth from [spiritual] darkness, and to whom it accreweth," Hobbes left the friends of liberty an unexpected gift. After reviewing the recent history of church-state relations in England, he wrote:

> [A]nd so we are reduced to the independancy of the primitive Christians, to follow Paul, or Cephas, or Apollos, every man as he liketh best: which, if it be without contention, and without measuring the doctrine of Christ, by our affection to the person of his minister, (the fault which the apostle reprehended in the Corinthians), is perhaps the best. First, because there ought to be no power over the consciences of men, but of the Word itself, working faith in every one, not always according to the purpose of them that plant and water, but of God himself, that giveth the increase. And secondly, because it is unreasonable in them, who teach there is such danger in every little error, to require of a man endued with reason of his own, to follow the reason of any other man, or of the most voices of any other men, which is little better than to venture his salvation at cross and pile. Nor ought those teachers to be displeased with this loss of their ancient

authority. For there is none should know better than they, that power is preserved by the same virtues by which it is acquired; that is to say, by wisdom, humility, clearness of doctrine, and sincerity of conversation; and not by suppression of the natural sciences, and of the morality of natural reason; nor by obscure language; nor by arrogating to themselves more knowledge than they make appear; nor by pious frauds; nor by such other faults, as in the pastors of God's Church are not only faults, but also scandals, apt to make men stumble one time or other upon the suppression of their authority.[21]

In short, forcing all subjects to follow their sovereigns' assorted religious beliefs would be a venture of one's salvation "at crosse and pile," that is, at heads or tails. And one should never leave such important matters to chance. John Locke would later develop an identical argument in his *Essay Concerning Toleration*, and carry it much further than Hobbes did, exploring its implications not only for religious belief but for religious practice as well.

Hobbes did not take this path. Coming at the very end of the lengthy *Leviathan*, his neglect seems almost forgivable. We should also remember the great radicalism of the idea, given its time. Religious toleration of a sort had come to England, not via a sustained philosophical argument, but by the simple failure of any unifying religious policy to win hegemony for very long. Hobbes had done no more than to point out that this was "perhaps" the best: Religious liberty meant that civil government would be less troubled by religious dissent, a result his contemporaries had good reason to welcome.

Not everything was praiseworthy, of course, in Hobbes' system: Hobbes admittedly reserved the censorship powers—among many others—to the sovereign, and this is no small matter. He also rejected all forms of checked or divided power, again no small matter. But in suggesting that essentially all economic, domestic, and educational functions might be left at liberty, and that the liberty of conscience might well be for the best, he justified a liberty that in some senses went beyond the liberty that most Europeans, then or now, enjoyed.

Finally, there was one other sense in which Hobbes should be counted a liberal. Although the initial grant of power in Hobbes' social contract was vast, permanent, and unchecked, we are assured by Hobbes that these powers were designed to be rare in their exercise: The sovereign would *have* all these powers, but he would have little cause to *use* them. Indeed, the very reason the sovereign would possess his power was so that power in general would be used relatively less often than it otherwise would. Hobbes' government was therefore designed toward a liberal end. The sovereign would hold power for the sake of *not* using it.

And, because there was only one sovereign, nobody else could get a comparable power, not even if they wanted it. Liberty would then be found not only in the silence of the law but also in the stillness of one's neighbors. Wrote Hobbes:

> [T]he sovereign power, whether placed in one man, as in monarchy, or in one assembly of men, as in popular, and aristocratical commonwealths, is as great, as possibly men can be imagined to make it. And though of so unlimited a power, men may fancy many evil consequences, yet the consequences of the want of it, which is perpetual war of every man against his neighbour, are much worse. The condition of man in this life shall never be without inconveniences; but there happeneth in no commonwealth any great inconvenience, but what proceeds from the subject's disobedience, and breach of those covenants, from which the commonwealth hath its being.[22]

But this was to praise the sovereign with faint damnation. It was to declare that a sovereign's government was a necessary evil, and that we endured it because its presence was "not so hurtful as the want of it."[23]

Hobbes' definition of liberty, finally, is not one that modern libertarians will (or should) recognize as their own. Although this book is not a history of liberty, some discussion of his peculiar understanding of liberty seems appropriate here. In *Leviathan*, Hobbes defined and rather consistently used the term "liberty" to mean the freedom from actual, physical impediments—a move Quentin Skinner terms "the most outrageous moment of effrontery in the whole of *Leviathan*."[24] It was certainly an odd definition. It clashed with the one from Scholastic philosophy—which held that freedom coincided with the capacity to do what a rational man would do—and it clashed also with the definition offered by the proto-libertarian Levellers, who, contemporaneously with Hobbes, held that civil liberty should be understood as one's actions not being dependent on the arbitrary will of others. If the Levellers were right, then liberty and law might coexist, provided only that the laws all were enacted with good reason, and were enforced impartially, and thus could not be called arbitrary. For Hobbes, however, liberty could exist only where law did not.

Hobbes therefore did not, and could not, believe that government served to preserve liberty. Government was strictly for the purpose of self-preservation. If, in the course of preserving people's lives, government happened to preserve some liberty, then that was so much the better, but—as his pre-*Leviathan* political works demonstrated—Hobbes was ready, if necessary, to jettison liberty altogether. In his 1640 work *The Elements of Law*, he had claimed, "Freedom therefore in commonwealths is nothing but the honour of equality of favour with other subjects … [I]n all other senses, liberty is the state of him that is not subject."[25] Civil liberty of any kind appears to have been a relatively late addition to Hobbes' thought.[26] As Skinner writes, Hobbesian subjects "retain almost no liberty whatever *as subjects.*"[27] When they are at liberty, it is because that liberty concerns a thing *outside of* the commonwealth. Skinner portrays Hobbes as having progressed from a more authoritarian to a considerably less authoritarian stance over the course of his life, and perhaps he is right, but if so, it is in large part because Hobbes had come to think that more and more things were rightly placed outside the commonwealth.

Hobbes' thought has had far-reaching consequences in political theory. If we hold that our states are artificial or conventional—neither natural nor divinely ordained—there would seem to be little other than the fear of bad consequences to impede us from tinkering with them. Our *rights*, meanwhile, may be recognized either as natural or as conventional, or perhaps as a mix of the two, without necessarily invalidating various proposals to refine the process of securing them. As a result, much of liberal political philosophy since Hobbes has tried to answer something like the following question: Can we find better means to the ends that Hobbes wanted, namely securing the prospects of peace and material security?

Another question is closely related to this one and also important in the history of liberalism: Shall the conventions that we set up simply provide a framework that allows us to *pursue* material security? Or shall our institutions attempt to provide material security more directly? To what extent, if any, does the attempt to provide material security more directly come to interfere with the framework of liberal political ends? These questions lead us to the political thought of John Locke.

LOCKE: GOVERNMENT IS A MEDIATOR

Much intrigue surrounds the writing of John Locke's *Two Treatises* and, in particular, the question of exactly when Locke wrote them. Ordinarily a question like this might not seem to matter so much, but for the *Two Treatises*, it does. Oddly, Locke appears to have destroyed all evidence of its drafting and all references to it in his personal papers.[28] The explanation for why he would do so bears surprisingly on our focal question.

Students often first encounter the mystery surrounding the *Two Treatises* in Peter Laslett's masterful introduction to the edition found in the Cambridge Texts in the History of Political Thought. The conventional dating of the *Treatises* had held, prior to Laslett, that they were written with a view toward justifying the Glorious Revolution and the accession of William and Mary in 1688. This was a reasonable theory, as the *Treatises* were published in 1690 and proposed to justify the overthrow of one monarch in favor of another. 1689 would have been a particularly safe time in history to offer such a theory; an overthrow of this type had just occurred, and it had been greeted with relief and celebration. James II, the openly Catholic king of England, had left the country. Parliament declared that he had abdicated, and it welcomed William and Mary in his place. Only a few minor skirmishes occurred, and the Glorious Revolution was done. The *Two Treatises* should fit right in.

Apparently, anyway. Following careful bibliographic study, Laslett concluded that the *Treatises* had not been written in 1688. A wealth of indirect evidence pointed to a much earlier date of composition for essentially all of the text. Laslett determined that it had been written around the time of the Exclusion Crisis, thus circa 1680, with the *Second Treatise* having been substantially completed *before* the *First*.

During the Exclusion Crisis, King Charles II had sought to keep his brother, the future king James II, in the line of succession to the crown, while a significant faction in Parliament sought to exclude him. Charles had dissolved three Parliaments in a row to prevent that. In this context, calling for the deposing of monarchs was a much bolder act. Rather than justifying a popular *fait accompli*, the treatises justified a revolution that had not yet occurred. Read in context, Locke becomes a radical.

But how radical was he? Locke was a close associate of Anthony Ashley Cooper, the First Earl of Shaftesbury, a man who was accustomed to making and unmaking kings. Shaftesbury had fought in England's civil war, first as a royalist, then as a parliamentarian, and ultimately he would serve on the parliamentary delegation that would invite the exiled Charles II back to England. Shaftesbury would later turn on Charles and plot his overthrow as well. Locke was probably aware of Shaftesbury's machinations and may even have been involved in them directly.[29]

To what degree this was the case remains a subject of scholarly dispute.[30] Yet Locke clearly moved in radical circles, and we should therefore consider suitably radical implications whenever they arise in his work. To the extent that Locke owed his livelihood or his chances for advancement to anyone, he owed them to Shaftesbury and not to the forces of the status quo, or to those who wielded (and sought to justify) arbitrary power. Locke may have been a new sort of political thinker in part because he had a new sort of patron.

Some aspects of Laslett's account have been challenged, however, including his ordering of the treatises' authorship.[31] Scholars today generally hold that the *First Treatise* was indeed written first, but that it was written in 1680. It was followed by a middle section that has mysteriously been lost, as Locke mentioned in the introduction. And then came the *Second Treatise*.[32]

But enough about the intrigue. Let's theorize—keeping in mind as we do that we are considering the ideas of a man who lived and breathed revolution, even if he didn't precisely foment one.

Much like Hobbes, Locke described people in the state of nature as having *"perfect Freedom* to order their Actions."[33] But Locke added that people were nonetheless bound by a moral law, one that preceded civilization and that was knowable to mankind by reason alone:

> The *State of Nature* has a Law of Nature to govern it, which obliges every one: And Reason, which is that Law, teaches all Mankind ... that being all equal and independent, no one ought to harm another in his Life, Health, Liberty, or Possessions ... [B]eing furnished with like Faculties, sharing all in one Community of Nature, there cannot be supposed any such *Subordination* among us, that may Authorize us to destroy one another ... Every one as he is *bound to preserve himself* ... so by the like reason when his own Preservation comes not in competition, ought he, as much as he can, *to preserve the rest of Mankind*, and may not unless it be to do Justice on an Offender, take away, or impair the life, or what tends to the Preservation of the Life, the Liberty, Health, Limb, or Goods of another.[34]

Moral laws of this type are admittedly peculiar: observers generally find them either plainly evident or plainly absurd. We should be mindful of the dissenters on this point, but mindful also that Locke was writing only for those who took his moral law seriously. Much remains to be fleshed out about it, including particularly how one legitimately comes by possessions, how best to preserve the rest of mankind, and how justice is to be done to offenders, but the *Second Treatise* at least addresses each of these points, if not always in ways that would satisfy us today.

We should add that this was a *moral* law, not a physical one, and the difference matters. Physical laws never need outside enforcement, but moral ones do. Without enforcement, a moral law only allows us to say when someone has acted well or badly. That's usually cold comfort to those who have been wronged. Worse, when enforcement is lacking, violations of a moral law proliferate. If we want to see *compliance* with what we take to be moral laws, we must have a means of enforcement. So Locke writes:

> And that all Men may be restrained from invading others Rights, and from doing hurt to one another, and the Law of Nature be observed ... the *Execution* of the Law of Nature is ... put into every Mans hands, whereby every one has a right to punish the transgressors of that Law to such a Degree, as may hinder its Violation ... [N]o Absolute or Arbitrary power, to use a Criminal when he has got him in his hands, according to the passionate heats, or boundless extravagancy of his own Will, but only to retribute to him, so far as calm reason and conscience dictates, what is proportionate to his Transgression.[35]

Transgressions of the Law of Nature may be justly punished. But not, Locke hastens to add, to any arbitrary degree; trespass is *not* to be met with torture or summary execution. If space permitted, we could multiply examples of various crimes meeting various punishments. We could debate their merits, and, with the benefit of practice and hindsight, we might elaborate a proper criminal code *even in the state of nature*. Locke doesn't do this, but the implication is clear: Government does not exist to establish the proper punishment for crimes; this proper punishment ought to be knowable to right reason even without government.

And yet already a problem has arisen: Almost foolishly, nature seems to have left every man as the judge in his own case. This will not do, because people are always biased toward their own interests. Locke writes:

> [I]t will be objected, That it is unreasonable for Men to be Judges in their own Cases, that Self-love will make Men partial to themselves and their Friends. And on the other side, that Ill Nature, Passion and Revenge will carry them too far in punishing others. And hence nothing but Confusion and Disorder will follow, and that therefore God hath certainly appointed Government to restrain the partiality and violence of Men ... *Civil Government* is the proper Remedy for the Inconveniences of the State of Nature, which must certainly be Great, where Men may be Judges in their own Case.[36]

The remedy is a plausible one, and a purpose for government has been declared. This is Locke's great innovation, namely, to propose that government exists to protect individuals' lives and property *in an impartial manner*. Government aims at objectivity in the enforcement of a moral law—a moral law that Locke understood to be universally attainable in theory, but never perfectly attained in practice.

In Locke's state of nature, there is certainly not an *undersupply* of property protection, which is to be rectified by creating a state. On the contrary, in Locke's state of nature, there is an *oversupply* of property protection, because each individual thinks too favorably of his own interests and acts too much with a bias in his own favor. Government is instituted to keep the supply of property protection under control.

Locke seems to have had a very particular notion of property in mind, and it is worth discussing it at some length. If, in answer to the question "What is government for?" we hear that "Government serves to protect property to the correct degree," we will have learned almost nothing if property itself is not clearly defined.

Peter Laslett stresses that property for Locke encompassed more than material goods and land. Property, writes Laslett, "is not defined as material possessions, nor in units of the conveniences or necessities of life ... [it] seems to symbolize rights in their concrete form, or perhaps rather to provide the tangible subject of an individual's powers and attitudes."[37] Thus it is not absurd to speak of having a property in one's mental faculties, or one's work-related skills, or one's family, with the latter not indicating any sort of chattel, but that one has a right to conduct family affairs in a manner free from outside interference. Laslett even notes how some of Locke's contemporaries used the same word to talk of having a property in the establishment of the Protestant religion.[38]

It seems clear that Locke himself intended property to encompass security in one's person, reliable assurances of fair dealings with the government and with one's neighbors, and, in extremis, even the right to self-defense. Laslett writes: "It is because they can be symbolized as property, something a man can conceive of as distinguishable from himself though a part of himself, that a man's attributes, such as his freedom, his equality, his power to execute the law of nature, can become the subject of his consent, the subject of any negotiation with his fellows."[39]

The ability to make these attributes of personhood subject to alteration by consent is, at long last, a path to *liberty under law*—to the exercise of free and reasoned choice that nonetheless need not conflict with living in a civil society among others. And for Locke, government was to concern itself with protecting these attributes of the individual, and with the administration of impartial laws for their disposition and enjoyment. In a chapter that recapitulates much of the *Second Treatise*, Locke wrote that people quit the state of nature, "which however free, is full of fears and continual dangers ... to unite for the mutual *Preservation* of their Lives, Liberties and Estates, which I call by the general Name, *Property*."[40]

It was for this reason that the exit from the state of nature was, in Locke's view, swift and certain. God had made a grant to mankind of the whole earth in common, but it could not be supposed that the commons would endure forever. God had intended the land to be used, and by the use of the land, title passed to the "Industrious and Rational," whose labor had increased the land's value. All that remained was to secure to them what was rightfully theirs, namely, the fruit of their labor. Little liberty was being given up, he thought, and what was surrendered was in any case not compatible with the freedom to be had in civil society.

Objections to Locke's system are numerous. For example: What about the right, which I formerly enjoyed when all property was held in common, to roam the world freely and in peace, never worrying about theft or trespass laws or about infringing on the so-called property rights of others? Hasn't the institution of private property in land and material goods, and its defense by a government, grossly wronged me? Locke and modern post-Lockeans like Robert Nozick have argued that the transition out of primitive communism and into a society of private property has itself provided compensation. As Locke wrote:

> [H]e who appropriates land to himself by his labour, does not lessen, but increase the common stock of mankind: for the provisions serving to the support of human life, produced by one acre of inclosed and cultivated land, are (to speak much within compass) ten times more than those which are yielded by an acre of land of an equal richness lying waste in common. And therefore he that incloses land, and has a greater plenty of the conveniencies of life from ten acres, than he could have from an hundred left to nature, may truly be said to give ninety acres to mankind: for his labour now supplies him with provisions out of ten acres, which were but the product of an hundred lying in common.[41]

Let's grant for the sake of argument that Locke was right about the productivity of enclosed farmland. What if I declare that I still prefer my right to roam freely? How does your ownership of productive farmland give you standing to dismiss my preference?

There are two answers to these questions, and I think that either one is sufficient.

The first is as follows. In rights theory, there may be said to exist two types of rights: *liberty rights*, which allow people permission to act in a particular way, and *claim rights*, which declare that certain actions of others must be either forbidden or required. These two types of rights are, when one considers carefully, two sides of the same coin: My liberty right not to be subject to arbitrary imprisonment, for example, also asserts a claim right on all other people, because they must refrain from arbitrarily imprisoning me.

The assertion of a liberty right to roam the whole world and take whatever one pleases also entails the assertion of a claim right, and a remarkably far-reaching one: The claim right is that all other people must refrain from all activities that might impede me. They must refrain from enclosing land to raise cattle, refrain from building houses that roamers might be excluded from, refrain from building rice paddies or hydroelectric dams, refrain from planting any hedgerows, and so on. Whereas refraining from imprisoning others is no great burden, the claim rights asserted here are unbelievably burdensome. They forbid any lifestyle other than being a hunter-gatherer. Further, it is by no means clear that the forbidden acts are blameworthy in themselves. In this, they are quite unlike arbitrary imprisonment. An argument might be constructed that leads to this conclusion, of course, but it cannot simply be asserted.

The second objection is that in the state of nature, and prior to enclosure, it may well be that no one would actually enjoy a right to roam and take. True, trespass and theft would not be crimes, legally speaking, but neither would arbitrary imprisonment or murder. And each person might expected, in Lockean terms, to be biased as a judge of their own case, overly severe toward strangers and overly forgiving of their own misdeeds. In a world of biased judgment, harshness toward strangers would come to seem well-grounded, and a vicious cycle would ensue. Roaming would almost never be safe.

Meanwhile, when faced with conflicting claims, use rights claimants in a governed state are constrained as to the means at their disposal. They may verbally protest. They may appeal to the authorities. And in emergencies, they are generally allowed only such means as may be justified after the fact. Excesses are to be punished. Insofar as governed states have failed in this respect, however, their failing does nothing to commend to us the state of nature, which will inevitably be worse. Roaming may be punished—or not—but any punishment is at least potentially limited by law.

It is therefore incorrect to complain that property laws in more or less the forms that we know them entail criminal punishments for trespass, and that these punishments infringe on liberty. It would much more accurate to say that property laws offer a possible *escape from punishment*—that is, an escape from the excessive, capricious, and all but inevitable punishments that are an everyday feature of the state of nature, whether before or after the enclosure of the worldwide commons.

The Lockean argument, however, cannot not be taken as justifying the holding of any particular piece of property, at least not without further evidence. Still less can we understand it as justifying the sum total of all holdings of property in any given system. Neither of these is a legitimate use of Locke's argument because it always remains possible that *no* property was in fact acquired through perfectly Lockean means, or that no adequate redress has been made. Over here in the real world, essentially all property has been acquired at one point or another through conquest.

This raises problems for strict Lockeans: Not only is the title to their property implicated but so too is their entire government, which owes its reason for being to the adjudication of a fundamentally illegitimate set of titles. One might argue that the state can be saved in that it also protects individuals' property in the larger sense that we have discussed earlier—it protects security of one's person, one's religious freedom, one's freedom from arbitrary punishment, and the like. This gets us quite far. But leaving all property in the conventional sense stuck in an undefended, possibly unowned state, without an impartial judge to dispose of it, seems hardly to fulfill the goals of Locke's system.

Patches abound. One modern-day Lockean, the libertarian economist and political theorist Murray Rothbard, proposed a serviceable one: Even when a property is known to have been stolen, "*if* the victim or his heirs cannot be found, *and if* the current possessor was not the actual criminal who stole the property, then title to that property belongs properly, justly, and ethically to its current possessor."[42]

Rothbard's patch solves one problem quite elegantly: In a world of broadly shared Lockean values, it will not do for large expanses of property to hang in limbo. Such property would be denied to the whole human race as far as productive use is concerned, while Lockeans would insist, after their progenitor, that land allocation rules, including those of private property, only exist so that land may actually be put to use, and not so that it may be sequestered forever. Property does not exist to serve as a sterile reproach to a set of ancient crimes.

We must ultimately resist the mistake of reading Locke's account of the origins of property and government as a literal historical description. Locke himself freely admitted that it was not, and David Hume insisted upon this point with particular force in his essay "Of the Original Contract." As with Locke, in Hume's philosophy property exists so that it may be put to good use, and it cannot be put to good use—or indeed, to any use—without some degree of stability of possession. Life in society imposes moral duties on all of us, Hume argued. Some of them are

> performed entirely from a sense of obligation, when we consider the necessities of human society, and the impossibility of supporting it, if these duties were neglected. It is thus *justice* or a regard to the property of others, *fidelity* or the observances of promises, become obligatory ... For as it is evident, that every man loves himself better than any other person, he is naturally impelled to extend his acquisitions as much as possible; and nothing can restrain him in this propensity, but reflection and experience, by which he learns the pernicious effects of that license...
>
> The case is precisely the same with the political or civil duty of *allegiance* ... Our primary instincts lead us, either to indulge ourselves in unlimited freedom, or to seek dominion over others: And it is reflection only, which engages us to sacrifice such strong passions to the interests of peace and public order.[43]

The Lockean state of nature is not to be understood as a literal prehistory of man. It is rather the counterfactual history of each of us, in the present day, if we were to consider ourselves as standing outside of society. That stance would entail many severe penalties for us, some deserved and others not, and it would inflict further penalties on others. It is from reflection on such a counterfactual that we come to appreciate that others hold their property with perhaps as much legitimacy as we would hold our own, and that an impartial judge is needed between us whenever conflicts arise. The social contract is not about the past. It is about today.

Will our attempt to create an impartial judge be perfectly successful? Of course not. Nonetheless it is hoped that we can all conclude that the society we inhabit is sufficiently restraining of the harmful passions of others, such that we hold our property, if not by perfectly unbroken chain of spotless title, then at least by sufficiently good title that we can exchange with others on a more or less equal footing, and that rectification of injustice is unlikely to disrupt our plans. We should hopefully all conclude as well that this society's judges have at their disposal sufficient procedures and resources to rectify improper transfers of property as they, and the appropriate aggrieved parties, are discovered. Such a process, though, is to be conducted against a background of security in property title. After all, this type of security is a kind of property in the greater sense, and one from which we all profit and depend upon: We can then be relatively sure that property we acquire will remain our own, and that's an assurance worth having.

This is also what Locke claimed that government was supposed to do: It was to allow all people the capacity, in principle, to acquire, improve, alienate, and consume various types of property, with the assurance that an impartial agent—the government—would judge disputes about the process, so that no one would be a judge in his own case, and that rules concerning the various transformations of property would be executed in an orderly and unbiased manner.

For Locke, that was *all* that government was to do: Government was not to *establish* property rights, for these rights were already established through a prior social convention. (Indeed, some recent research suggests that this is still a viable method of property allocation today.[44]) Nor was government to *protect* property, except insofar as the government's very presence would deter crimes against it. Further, government was not to order society or instantiate any divinely ordained plan for it; indeed, this was a purpose of a type Locke spent much of the *First Treatise* ridiculing. Government was not to display a unitary will at all—on the contrary, to judge the cases of everyone impartially seems almost to require the lack of a particular will, the incapacity to impose a pattern that is declared in advance. A pattern of that type would inevitably be partial to one party or another.

In describing government's purpose in this manner, Locke achieved a significant intellectual feat: Government was no longer described as the natural completion of the self, or the link between the self and the Divine, as many had claimed. Nor was it simply a curse, as others had claimed. Locke had characterized government as having a specific set of tasks before it, tasks which it could do either well or badly. In the Lockean paradigm, one could then evaluate the government's performance in ways that might—or might not—lead to a rejection of the social contract as it had been carried out. Lockean politics rejected totalization in favor of a backdrop of continuity and conventional order, one in which most ordinary citizens, in most occasions, could go about their lives and *not* have to concern themselves much about their relationship to the state.

Politics had been defanged, at least in theory. Locke's theory of government would prove highly influential, so much so that to some degree, every significant author we will discuss in the following pages wrote in response to him. Locke didn't provide the first justification for private property, but one might easily get the impression that he did. Locke was nowhere near being the first social contract theorist, but his social contract remains the one that we tend to silently assume when discussing the idea. And Locke did not provide the first justification for the right of revolution, but he unquestionably supplied a very effective one that is still sometimes invoked today. Perhaps Locke's greatest radicalism, though, lies in his answer to the question of what government is for. And on this point, very few have followed his lead: Government exists to settle disputes impartially, and for no other reason.

It is interesting to contemplate what a government might look like if it were much more thoroughly Lockean than our own. If government existed *only* so that individuals would never serve as judges in their own cases, then the entire class of crimes commonly termed victimless would have to disappear. The theory of punishment would also have to undergo a radical transformation: If the purpose of government were merely to resolve disputes, then both retributive punishment and punishment that aimed to rehabilitate the criminal would have no justification whatsoever. Only restorative or compensatory punishments would be justified. Nothing about this government's purpose would seem to demand an exclusive geographic jurisdiction, as modern states like to claim, provided only that for all cases, there were some means of establishing a unique dispute resolution venue. Disputants might enjoy wide leeway in choosing an arbiter agreeable to them both, based on the arbiter's reputation for fairness and/or expertise in the area of dispute. The whole would perhaps resemble the system of private protection agencies outlined in David Friedman's *The Machinery of Freedom*, although this is undoubtedly a more radical conclusion than Locke would have accepted.[45]

ROUSSEAU: THE SEARCH FOR LOST VIRTUE

Jean-Jacques Rousseau was a complex thinker, particularly regarding our focal question, to which he supplied not one but two distinct answers: Governments as they existed in the world were an affliction. They entrenched material inequality, fomented social antagonism, and actively corroded civic virtue, all so that the few at the top could pride themselves on having more status and possessions than the rest. The existing governments were more or less uniformly evil. Yet governments founded according to Rousseau's own principles would serve a different purpose; they would exist to protect virtue from corruption. Were these ideal governments possible? On that question, I believe that Rousseau was pessimistic; he thought it unlikely that anyone could found such a government. Even if they did, the ideal government would not last for very long; the rest of the world would soon destroy it.[46] Nonetheless Rousseau's portrait of a virtuous government has attracted many admirers, and certainly too many would-be practitioners for our study to ignore.

Rousseau is a rich feast, too; it's easy to see why he appeals to intellectuals. In his writings, he commonly presented multiple viewpoints, often with some degree of sympathy, and he alluded to many others. He was frequently ambivalent and nuanced about mankind's moral psychology, including particularly our prospects for happiness, our ability to exercise sound judgment, and our various forms of social organization. Rousseau makes for engrossing and often compelling reading. His deployment of the modern political idiom, including concepts like liberty, social contract, and social class, makes him among the most contemporary-sounding of eighteenth-century writers.

Perhaps the biggest paradox about Rousseau, then, is that in the end he settled on a return to the past: His ideal government was essentially that of Sparta. He argued for the ancients and for the reign of ancient virtue even as he grappled with distinctively modern problems and used modern political language to do it.

Through Rousseau, communitarianism of the ancient type was in effect reborn. And here is a further paradox: His intellectual children include modern populism, modern nationalism, and even modern environmentalism. These ideologies all assign many important tasks to the state—and not just to some projected or utopian state, but to the actually existing one. Rousseau would certainly not have wanted any such thing. He would not even have wanted to instantiate the system described in his most famous work, *The Social Contract*, for the simple reason that he did not think any contemporary population was virtuous enough to do it.[47]

All existing governments were to some degree corrupt, Rousseau held, and their corruption was the product of an even deeper set of social and moral corruptions. These were rooted in human psychology, which exerted a controlling influence over all political institutions. The remedy to existing evils was not to be found in the forms of government, which could provide only a temporary check on human corruption at best. No, healing the human condition required nothing less than the full regeneration of morals, and the renunciation of modernity as Rousseau (and we) understand it.[48]

This disease of civilization had an unlikely source: Rousseau believed the progress of the arts and sciences was to blame. True, such progress had produced much that was at least superficially pleasing, but Rousseau held that this *material* progress had yielded no *moral* progress at all. Rather the opposite; progress in the arts and sciences was in fact the root of modern man's discontent. Misery in the midst of abundance was therefore easy to explain: Abundance *causes* misery.

To Rousseau, there were two essential problems with material progress. First, it made humanity generally softer and weaker, both in body and in spirit. And second, the fruits of the arts and sciences, and of civilization, would always be unequally distributed. Both of these had far-reaching negative consequences, particularly the latter. Inequality was perhaps the central fact about civilized life to Rousseau. It was also mankind's greatest tragedy.

The disaster of inequality had been unfolding all through human history. Primitive man had undeniably been both stronger and happier, claimed Rousseau. Before civilization, he pictured man "satisfying his hunger at the first oak, and slaking his thirst at the first brook; finding his bed at the foot of the tree which afforded him a repast; and, with that, all his wants supplied."[49] Unlike Locke, Rousseau's state of nature wasn't a counterfactual or a thought experiment. It was a specific era of time, namely prehistory. At least if we credit Rousseau's account, prehistory put modernity to shame, not materially, but morally.

If the accommodations in the state of nature seemed poor to Rousseau's readers, it only went to show how far the readers themselves had fallen: Acorns and water were perfectly adequate for the virtuous man of nature. Only soft, effeminate, *civilized* man could have any cause for complaint.[50] Not only were natural men constitutionally stronger, they were happier— and freer—as well:

So long as men remained content with their rustic huts, so long as they were satisfied with clothes made of the skins of animals and.... . confined themselves to such arts as did not require the joint labour of several hands, they lived free, healthy, honest and happy lives ... But from the moment one man began to stand in need of the help of another; from the moment it appeared advantageous to any one man to have enough provisions for two, equality disappeared, property was introduced, work became indispensable, and ... slavery and misery were soon seen to germinate and grow up with the crops.[51]

The division of labor among many people, which allows much a more complex material culture, is the hook that ultimately snares us. Anthropologists in the modern era have long noted, and lamented, the near-universal tendency of less-developed peoples to take up modern ways of life as soon as they become available. Deviations are not unheard of, but they achieve notice in part because they are remarkable. This tendency suggests, perhaps embarrassingly, that Rousseau's happy noble savages were either not actually happy—because their revealed preferences say otherwise—or were not actually noble—because they give in so readily to temptation.

Modern life, with its fresh-baked bread, its warm-water showers, and its fiberglass insulation, has after all a certain appeal to it. But the Rousseauian response to all this, which anthropologists have sometimes adopted, is to claim that, yes, humans are too easily corrupted.[52] We moderns, meanwhile, are in no position to judge ourselves better than our uncivilized, uncooperative ancestors; in Rousseau's system, a one-way street runs from virtue to corruption, and we stand at the wrong end of it. Worse, it is only with the greatest possible difficulty that any one person succeeds at returning to virtue. The regeneration of a whole society is essentially never to be expected.

Attentive readers will note that this account rests on some exceptionally strong empirical claims in both natural history and moral psychology. Rousseau did virtually nothing to support them. He was no anthropologist, and his empirical evidence about the lives of people in the state of nature seems to have been essentially nil.

He appears, rather, to have developed much of his understanding of what "nature" was by a sheer process of introspection. As Judith Shklar writes, Rousseau believed that "he was not like other men ... nature still spoke through him and he was willing to reveal it to others." He had "[a] soul still natural, the will to look into himself and to display what he saw and unusually rich social experiences," Shklar continues.[53] In short, he *intuited* what life in the state of nature must have been like, and he held to that intuition with a lifelong firmness. Nor was Rousseau deterred by the fact that he was personally incapable of living the life of the natural man. On the contrary, he attempted to use his inability to live in the state of nature as further proof of his system: He may have had an uncorrupted soul, he claimed, but his body had been civilized and made weak, more or less like everyone else's. A convenient distinction: As historian Mary K. McAlpin has observed, Rousseau carried the exaltation of his natural soul, and the apology for his corrupt and feeble body, even into his sex life. His well-known urinary ailment was no venereal disease, Rousseau claimed; rather it was, like so much else, the product of an enfeebling civilization.[54] By this compromise, Rousseau could enjoy the comforts of civilization while constantly condemning them, thanks to his claimed insight into the state of nature. He did both in abundance.

In Rousseau's moralized history of mankind, the introduction of "iron and corn" was the fatal step "that first civilized men, and ruined humanity." Iron made the defense of private property feasible; corn made enclosing the land profitable. Civilization began, Rousseau famously claimed, when a man first claimed a parcel of land as his own. Soon there came other innovations, things like language, family, farm implements, and constructed shelters. These, the products of human ingenuity, made primitive man's life materially easier. Food became more plentiful, and love became possible in something like the sense that we now understand it.

It was all downhill from there. The new inventions took away the roughness and the strength of primitive life. Acorns just weren't enough anymore. Formerly simple shelters grew better and better, and woe to the ones whose shelters weren't in the latest fashion. Even love now competed, fatally, with lust, which hadn't gone away. Whichever of the two one chose, the other would always beckon, and it would always leave the chooser unhappy.

Time turned everything sour, including all of the new inventions: "these conveniences lost with use almost all their power to please, and even degenerated into real needs, till the want of them became far more disagreeable than the possession of them had been pleasant. Men would have been unhappy at the loss of them, though the possession did not make them happy."[55] Many subsequent critiques of modernity have gone no farther than Rousseau's essential framework in their moral psychology, and yet critiques of this type are perennially popular, even among readers too soft and delicate to bear a return to the eighteenth century, let alone to the state of nature.

With new inventions, inequality likewise grew over time: With more and more wealth to go around, the distribution was necessarily less and less equal. One might think that this unequal distribution wouldn't matter so much, particularly if those at the top were not actually much happier. But inequality led to envy—and thus to misery, even among those who in absolute terms had greater material holdings than natural men ever had. The *labor* of civilization, the efforts of its hardest workers and its greatest intellects, was therefore in vain. In civilization, resentment flourished among the lower classes, while vanity, cruelty, and indifference to human suffering flourished among their supposed betters.

Rousseau believed that envy and inordinate self-regard—what he called *amour-propre*—had corrupted everything in the civilization that he beheld around him. He praised the simple peasant life in contrast to the corrupt life of the cities, and particularly that of Paris. Yet even the present-day peasants were in danger, he argued, of losing their innocence.

As with his account of natural man, data played almost no role in Rousseau's claims about peasants. Instead Rousseau relied almost exclusively on an examination of his own experiences and his feelings about them. Rousseau, writes Shklar, "never wrote about the actual life of either the urban or rural poor ... His peasants are prosperous utopian figures designed to shame the degenerate civilizees of Paris." An ethnographer he certainly was not, even by the relatively loose standards of the time.[56]

His own life became the lens through which he read the operation of class, morals, sensibility, and the inner lives of others. "Living among the opulent," he wrote in one passage of his *Confessions*, "but without being able to keep a household as they did, I had to imitate them in many things, and what were for them little expenses, were ruinous for me." In the time he wrote of, Rousseau was already a successful author, but he still couldn't spend money like upper nobility could. Someone else, perhaps, might not have felt the same obligations, but this does not occur to him. Having no footmen, he had to do all the work of keeping up appearances—work which he evidently thought important, but which he resented—entirely by himself. When his social superiors tried to help, they made matters worse through their imprudence with money. To top it all off, Rousseau found no pleasure at all in high society. "If this lifestyle had been to my liking," he wrote, he might have tolerated it. "But to ruin myself for the sake of being bored is insufferable," he declared.[57] Undoubtedly it was, for him. But whether this experience can stand as evidence on which to base a critique of an entire society is another matter.

Simply being around *les grands* was draining to Rousseau, both financially and spiritually. He had been to the top, he declared, and it wasn't worth it. And, he inferred, *absolutely nothing in civilization ever was.* What Rousseau had felt to be true in his own life became true of society at large, and *no one,* he ultimately believed, was really made happier in any lasting way by any form of material comfort. The state of nature simply had to be better.

In any case, Rousseau held that comfort was frequently not the true purpose of material progress. Almost by definition, luxury goods could go only to a few; a luxury would hardly be worth the name if the many could possess it. That's because the entire point of having a luxury was not to enjoy it, but simply to flaunt it before one's inferiors. In Shklar's words, the wealthy sought luxury "in order to be admired and to be able to look down upon the less fortunate ... inequality is the whole object of their striving."[58] And this striving—ceaseless, remorseless, and unsatisfying—*this* was the stuff of civilization. It promised happiness if one could climb the social ladder. But very few could do that, and when they did, they found themselves on what we might today call a hedonic treadmill: They would have worked a great deal, but they would have made no progress at solving the problem of envy. The joy they felt from flaunting what riches they had would always be fainter and shorter-lived than the envy they felt from those above them. Like the primitives who had invented handheld farm implements, they got a one-time rush of happiness, but a lifetime of slavery for themselves and their descendants.

And so the peasants would envy the bourgeoisie; the apprentices would envy the masters; the masters would envy the nobility; the nobility of every rank would envy those whose rank was just slightly above their own. Only one man could be king, and in any case, Rousseau did not believe that kings often led happy or virtuous lives. "We see around us hardly a creature in civil society," he wrote, "who does not lament his existence.... I ask, if it was ever known that a savage took it into his head, when at liberty, to complain of life or to make away with himself."[59]

This antagonistic, fundamentally unhappy picture of civil society has had a long and storied career, one that continues in the present day. When modern liberals condemn classical liberals for being atomistic individualists, this condemnation owes its origins to Rousseau, and to modern liberals' own picture of civil society, which also comes from him. Like Rousseau, modern liberals often regard civil society as fundamentally atomistic and hostile, without any real possibility of mutual benefit or coordination, unless it is imposed from above, a task that they look to government to perform. They believe that they, the modern liberals, are well-situated to guide the government in this task, while the classical liberals resist it. That is, mutual benefit and coordination would appear to come only from the government, and never simply from a voluntary action of parties whose interests are either aligned with or else not in material conflict with those of others.

Whenever a government ruled over a fundamentally bad civil society, Rousseau held that its purpose was to entrench everyone still further in the whole sordid system. The very origin of law, Rousseau claimed, may have been when a group of rich men devised a cunning plan to guard their own riches, which they named government, and which was "as favorable to themselves as the law of nature [was] unfavorable."[60] From then on, "All ran headlong to their chains, in hopes of securing their liberty; for they had just wit enough to perceive the advantages of political institutions, without experience enough to enable them to foresee the dangers … Even the most prudent judged it not inexpedient to sacrifice one part of their freedom to ensure the rest; as a wounded man has his arm cut off to save the rest of his body."[61]

If only we had *not* listened to reason, but rather to the heart, and to wise people like Rousseau! But how might one construct a society that avoided this appalling mess? As we have mentioned, Rousseau did not think that it could be made of the stuff of the ordinary societies around him. It could definitely be sketched, though, and the purpose of the sketch would be to shame all those in contemporary society who still had a conscience.

Rousseau's use of modern language to further an ancient cause can be seen in the title of his most important political work, *The Social Contract, or Principles of Political Right (Du Contrat social, ou Principes du droit politique)*. As we have seen, the term "contract" and its close synonyms had become nearly ubiquitous in the early modern era as a way of describing the act by which societies instituted government. Certain of these terms, particularly "contract" and "agreement," carried special implications; they suggested that the act was, first, one of *convention*, not of nature, and, second, that the act's terms were *limited in principle*: Whereas a covenant might (and, in religious history, did) encompass all aspects of life, a contract or an agreement generally speaking did not. As we have seen, both Locke and Hobbes proposed social contracts in this limited sense.

The same cannot be said of Rousseau. His social contract was not limited in principle. In part, this was because it did not create a *government*; it created a *society*—in all of its aspects. Rousseau could not possibly have been clearer on this point; indeed, he devoted Book III, chapter xvi of *The Social Contract* to the proposition "that the institution of government is not a contract."

The institution of society, however, was a contract, at least for Rousseau, and in contravention to how the term had formerly been used. Rousseau's social contract began at the moment when a set of individuals had come to think of itself as a permanent and indivisible collectivity; only afterward could that collectivity sensibly designate a subset of itself as a government. Without a preexisting sense of total unity, no proper state could exist. Even a roughly Lockean government, wrote Rousseau—that is, one that "will defend the person and goods of each member with the collective force of all"—could not occur without achieving complete social unity. To have a proper society, one needed

> the total alienation by each associate of himself and all his rights to the whole community. Thus, in the first place, as every individual gives himself absolutely, the conditions are the same for all, and precisely because they are the same for all, it is in no one's interest to make the conditions onerous for others.[62]

Two confusions are likely to arise here. First, how can one create a government that will protect one's claims to one's person and goods—just *after* one has created a society, to which one has surrendered all rights to one's person and goods? Rousseau is fairly explicit about the surrender, too, writing, "since the alienation is unconditional, the union is as perfect as it can be, and no individual associate has any longer any rights to claim."

Rousseau solved this first problem by arguing that following the surrender of all rights to society, a good government would return all things to each who had surrendered them: The *natural right* of possession dies, but the *legal right* to property is born, thanks to the government. The latter resembles but is unrelated to the former.

The second confusion now arises: The rights surrendered in Rousseau's version of the social contract were *not* surrendered to the state, which did not yet exist, but rather to *society*. By what process or title does the state come to have them? Simple, replied Rousseau: Society's next acts are to create a state, to perform the requisite transfers, and to instruct the state on how to govern from then on.

Rousseau thus stands at the origin of the modern notion that whatever rights we may have are received not from God or from nature, but from the state. This is not a claim that I find persuasive. Although it is true that the exercise of rights not recognized by the government is a less than secure endeavor, this is not the same thing as saying that only government-recognized rights are philosophically defensible. Many things may be philosophically defensible in the abstract, while unwise or impractical in a given time or place.

Applied consistently, the principle that the state is the source of all rights would deprive its adherents of the ability to criticize any government policy whatsoever. That which the state asserted in the way of rights claims would be ipso facto just, no matter how ill-conceived or iniquitous we might otherwise wish to find it. The principle fixing the source of rights in the state may have begun with the aim of justifying the rights of man to the skeptical; it ends, though, in moral nihilism.

A further difficulty to Rousseau's social contract, which Rousseau does not seem to have sufficiently considered, is that even if the rights of man in a well-governed society are "the same for all," it is unclear why individuals will strive to keep it that way thereafter. Merely because my social conditions are the same as yours, how does it come to be that I escape the thrall of envy? Won't it still be there, in that I can always still imagine being wealthier?

More innocently, it is unclear why ordinary political differences will not continue to arise in a good polity, why they will not sometimes be deep and disuniting, and why citizens of a good polity are in such cases legitimately obliged to obey the government at hand rather than their own consciences. Some commentators on Rousseau, notably Joshua Cohen, have argued that Rousseau in effect solved the problem of political legitimacy by positing that a political society *only* exists in the context of a pre-political agreement on what constitutes the general good for the society at hand. Such an agreement would certainly forestall future political conflicts, at least among the sufficiently virtuous. "A free community of equals requires a shared understanding of and allegiance to the common good," Cohen writes.[63]

And yet one does not solve the problem of political legitimacy by limiting one's discussion to those communities in which the problem of political legitimacy never arises. Such a move may appear to win the game, but only by never actually playing it. Moreover, even a successful agreement on the nature of the common good will require continuous re-instantiation, in the form of real-world institutions, from one generation to the next, in the face of changing political and social forces that the initial contracting parties cannot anticipate. In a very real sense, the passage of historical time renders the problem of political legitimacy immediate to every age. A pre-political social contract hidden in the mists of time will not suffice to solve it. It is fine-sounding to write, as Cohen does, that "sovereign authority lies in effect in the shared understanding of the common advantage," but shared understandings do not govern, and sometimes they are not even shared.[64]

To be fair, Rousseau was indeed pessimistic about the odds of a good government lasting long in the world. Nonetheless, he held that one could only achieve freedom *as a citizen*, and only as a citizen of a good government. And his pessimism went further still, because on those rare occasions when freedom did arrive, the citizen would often fail to recognize it. Rousseau wrote:

> The citizen consents to all the laws, even to those that are passed against his will, and even to those which punish him ... The constant will of all the members of the state is the general will; it is through it that they are citizens and free ... When, therefore, the opinion contrary to my own prevails, this proves only that I have made a mistake, and that what I believed to be the general will was not so. If my particular opinion had prevailed against the general will, I should have done something other than what I had willed, and then I should not have been free.[65]

In a footnote, Rousseau observed approvingly that the authorities in Genoa inscribed the word *Libertas* on the doors of the prisons and the fetters of the galleys. All that is individual rather than collective Rousseau terms unfree; all that is collective he calls free, even the prisons, which are simply where the unvirtuous discover freedom.

The reason for this strange identification of freedom with prison is that, for Rousseau, there was a sense in which our particular will is not our own. Modern man desires luxuries, but such desires are enslaving; indeed, they corrupt our very wills. The ideal state would preserve its citizens from corrupt desires and preserve their authentic wills in the process. Most of us, possessed of inauthentic wills, cannot attain such heights. A really good prison might help.

At this point in Rousseau's argument, he may stand open to the charge that he has committed a No True Scotsman fallacy, in which an arguer modifies the subject of a proposition after the fact, so as to render the proposition true. It runs something like this:

"Good citizens cannot will anything other than the general will."
"I'm a good citizen, and I will something other than the general will, because the general will is an illusion."
"Welllll … good citizens can't *authentically* will something that is other than the general will."

But in a sense the adherents of Rousseau's system were condemned already: They would have conceded their lack of authenticity some time ago; insofar as they were not willing to subsist on acorns and sleep under trees, they lacked the naturalness that was the only basis of authenticity. Whether true or not, conceding the point of one's own corruption would preserve the role of the state as the guardian of virtue. Objections from moderns are overruled, and we are, and remain, moderns.

It would not be far off the mark to call this a philosophical con game. The aim would seem to be to remove the victim's confidence in all things other than Rousseau's philosophy; the trick's genius consists of beginning with innocuous and commonplace observations—that sleeping in a feather bed in a warm house is preferable to sleeping under a tree, or that bread is preferable to acorns—and then impugning them. Only a few—the few with the most natural and uncorrupted souls—would be capable of realizing that comforts were marks of corruption. Rousseau, of course, was just such a noble soul. He himself had said so. Even if his own physical life was far from Spartan, he could *feel* that something was wrong with it. Now answer: Are you with the few who can feel that something is wrong? Or are you with the many who are corrupted?

As the Polish communist-era dissident historian Leszek Kołakowski asked in a related context: "Is a man really entitled to claim that everyone is out of step except himself?" Are we really to believe that all of the institutions of civilized life exist solely for wickedness, and that somehow no one ever recognized it until Rousseau? How does one respond to such an audacious claim? Contemporary readers knew exactly how to respond, and they did so with enthusiasm: They too were pure of soul, they insisted, or at least they were hoping for a regeneration. Rousseau became an idol to his contemporaries.[66]

Rousseau's pessimism notwithstanding, both contemporaries and later actors eagerly pressed his thinking into the service of various real-world governments, most notably in revolutionary France. Indeed, Rousseau's influence can be seen everywhere in the French Revolution, up to and including the famous invocation of the General Will in the Declaration of the Rights of Man and Citizen. Even when applied to unfit objects, which France certainly was for Rousseau, his thinking has stirred the hearts of millions toward specific, real-world political acts.

Unfortunately, Rousseau's philosophy makes an excellent fit to a would-be tyrant's purposes: As we have just seen, the Rousseauan state is not intellectually open to critique; critiques from any quarter may always be deemed inauthentic and therefore worthless. Yet the Rousseauan state may always be subject to modification from within, perhaps arbitrarily, by those who already wield authority. After all, none may criticize. With its direct assault on the both man's reason and on man's material progress, Rousseau's ideology is ultimately corrosive of many conventional forms of sociability.

Rousseau was in a way a kind of Platonic statesman—and not simply a statesman of the government, but a statesman of the human heart and of social relations more generally. He, and he alone, had somehow stood outside of the common run of experience, which had blinded everyone else. He had grasped how the whole system worked, and now he was going to explain it all to you, the reader. It is true that Plato staked his authority on reason, while Rousseau deployed sentiment and naturalness. They nonetheless shared much in common. Both thinkers themselves seem to have been far more content to create polities in the mind or in words than to create them in the real world. Both held a frank disdain for the messiness of the real world, and for material goods in general. Both were profoundly skeptical of commerce. And the acolytes of both would find their work useful in creating totalizing states: The sheer existence of an authority who had seen to the bottom of it all would seem to demand such action.

We have encountered this type of self-aggrandizing social theorist several times already, and we will encounter it several times again. It is as if the urge to rule over one's fellow man in the physical world has been sublimated: It has become an urge to declare oneself the king of ideas and words. No doubt, it is exhilarating to imagine oneself in this role, and it is maybe only a bit less exhilarating to read someone who can convincingly project the image. Yet as previously observed, exhilaration is not a reliable guide to social theory.

As a result, much of what one thinks of Rousseau will depend on whether one feels, with him, that civilization and material progress aim at no other purpose than the conspicuous display of inequality, or that envy arises from civilization and is both insuperable and all-corrupting.

Not all have agreed. In his landmark book *A Theory of Justice*, twentieth-century philosopher John Rawls held that material inequalities, even great ones, might be tolerated whenever they redounded to the absolute benefit of the worst-off social class. That is, if the gap between rich and poor were to grow somewhat, but if the poor were nonetheless better off than they were before the changeover, the society could be said to have improved, provided that rights of political participation and other basic non-economic liberties were not compromised.

Rousseau would not have agreed. For him, justice wasn't about bean-counting. It was about returning to a simpler time, if one possibly could. Failing that, it was simply about denouncing the present, as loudly and forcefully as possible, and in the name of a particularly antique mode of virtue.

KANT: GOVERNMENT IS AN UNFINISHED PROJECT

Immanuel Kant was above all a champion of free inquiry and of the power of human reason. Contrary to common misconceptions, although he identified certain topics about which he believed reason must remain silent, in other cases he affirmed the power of reason to understand important truths. Kant was also an ethical individualist. He supported free trade, private property, and an objective standard for right and wrong conduct. He looked forward to a future of ever-improving legal regimes that would more and more respect the autonomy and dignity of every human being, and he urged all nations toward a just peace with one another. In short, Kant was what we might term a classical liberal.[67] His theory of government is crucial to our study for two main reasons: First, he exerted an extraordinary influence on subsequent philosophers of many different types, as we will see in later chapters. And second, Kant declined to describe his ideal state in great detail. He placed the perfection of civil society at a distant, indefinite time in the future, and this move freed him from having to make precise claims about perfected technologies of government. Kant limited himself to writing about what he thought must necessarily be true about an ideal state—while leaving much to be worked out later and confessing that a good deal about the perfected society must remain unknown. Kant deliberately declined, then, to build a city in words, as Plato had done. By placing his ideal city in the future and disclaiming full knowledge of it, Kant's system remained explicitly open to development and growth, and it is a mark of his philosophical depth that he left the door open for others to improve on his thought.

At least a brief look at Kant's ethics is necessary before considering his views on government. How can we ever find common ground in ethics, he asked, not just with some people, but with all people, both now and in the future, idealized society? What is the thing toward which all other good things are tending? The answer, Kant said, was a good will. The cultivation of a good will, and the subsuming of all other desires to the development of a good will, was for Kant *the* work of ethics, the end toward which all other ends pointed. Conscious, rational inquiry about the good, and the pursuit thereof, was *in itself* the highest good we might have, and all acts that manifested a good will were for this reason to be counted good as well.

Crucially, having and acting on a good will requires *autonomy*, a word that Kant used to denote our capacity to set forth ethical rules for ourselves and to follow the rules we had set forth. A fundamental ethical law would therefore have at least three important attributes: It would be objective in nature, and thus not subject to arbitrary whims or desires; it would be based on reason alone, and thus intelligible to all ethical agents; and, finally, it would be of a type that we could freely but deliberately subject ourselves to it. As Kant wrote in the *Groundwork of the Metaphysics of Morals*, "the basis of obligation must not be sought in the nature of man, or in the circumstances in the world in which he is placed, but *a priori* simply in the conception of pure reason."[68] A commitment to reason itself should form the groundwork of ethics.

Beginning with the necessity that reason must not contradict itself, Kant arrived at what would come to be known as the first formulation of the *categorical imperative*—called "categorical" because it was to apply to all ethical agents, in all circumstances whatsoever. It ran as follows:

Act as if the maxim of your action were to become by your will a universal law of nature.

As with all reasoned laws, the moral law must be consistent. We must be able to will that its maxims should be enacted *for all moral agents*. If we cannot do so, then we should reject whatever maxim we are considering.

For example, I cannot reasonably will that all people should steal, for this maxim cannot be universalized consistently. As a maxim, "commit theft" requires the willing agent *both* to believe in the justice of a given property holding *and* to believe in the justice of violating it. It is not merely that a world full of thieves would be a miserable place, although certainly it would be. The deeper problem is that I cannot consistently will *both* the existence of legitimate ownership *and* its occasional, *ad hoc* violation. Many similar principles of conduct are likewise ruled out, as reflective readers will quickly appreciate.

The first formulation can thus be understood as a kind of test for moral maxims. These stand to be refined over time as they are reconciled with one another and increasingly clarified. It may seem, however, that the first formulation gives little clear guidance about the nature and purpose of government. Kant's second formulation of the categorical imperative may help us a bit more:

> Act so that you use humanity, whether in your own person or in the person of any other, always at the same time as an end, never merely as a means.

Closely related to this second formulation is his third formulation:

> Every rational being must so act as if he were through his maxim always a legislating member in the universal kingdom of ends.

Kant claimed that these three formulations were all restatements of each other, unfortunately without explaining what he meant by this. Yet one way of thinking about it may be simply that our *maxims* will always sooner or later implicate *rational beings*—at the very least, they will implicate ourselves—and, as a result, they must always proceed from a correct understanding of the attributes of rational beings. If our maxims fail in this regard, they will be *inconsistent*, and thus impossible to universalize. As we honor the ethical search in ourselves, we must do likewise for others; we must recognize that they are on a similar quest to our own. That quest requires us to seek and be bounded by universality, and to recognize that all other rational beings should do the same. We are, then, all to consider one another as legislating members in a universal kingdom of ends.

The political implications now come into sharper focus. In common with Aristotle, Kant held that the search for the good is an end in itself, regardless of where one might find oneself in the search. Again like Aristotle, Kant held that people are fit for social interaction, but, he held, we must be guided by the categorical imperative, and we must not use any ethical seeker merely as a tool for our own purposes. Here, then, is the ethical basis for the *commonality* of human dignity, and the justification for those laws—and only those laws—that treat individuals with an equal initial respect. Aristotle's *polis*, the small community of citizens surrounded by a large population of slaves, will not do, not even if the end is the political life of the city.

Indeed, many social institutions, and perhaps most of them, will also fail the test. Plausibly the ideal polity would never be permitted to order individuals to build a bridge, or go to war, or even pay taxes, although Kant personally believed that it could. Still, one might argue that compulsory taxation constitutes a violation of the categorical imperative, because it treats the citizens merely as a means toward a greater end, the end desired by government planners, on which the tax money would be spent. Following Kant, twentieth-century libertarian philosopher Robert Nozick would later take a similar path:

> [M]oral side constraints upon what we may do, I claim, reflect the fact of our separate existences. They reflect the fact that no moral balancing act can take place among us; there is no moral outweighing of one of our lives by others so as to lead to a greater overall *social* good. There is no justified sacrifice of some of us for others. This root idea, namely, that there are different individuals with separate lives and so no one may be sacrificed for others, underlies the existence of moral side constraints, but it also, I believe, leads to a libertarian side constraint that prohibits aggression against another.[69]

Actually existing governments, however, *routinely* break this side constraint, in Kant's time, in Nozick's, and in our own. Quite possibly governments are incapable of doing otherwise. To speak more precisely, in the modern world, *the agents of states* set goals that they wish to attain, and they compel citizens to try to attain them, the citizens' dignity and autonomy notwithstanding. States as we know them—states with purposes independent of the wills of their constituents — therefore stand under a severe moral indictment. Far from commanding our respect, most political leaders should be held in contempt. It would seem that the entire point of their existence is to treat people merely as tools, and not as moral agents.

That's at least potentially a big problem for the legitimacy of the state as we now understand it. Yet Kant believed that it was nonetheless rational to enter into a civil society, with a government, and even an imperfect one, for two reasons. First, this situation was generally preferable to the ungoverned state; and second, it was only in the context of civil society that mankind could pursue the project of the gradual perfection of human institutions.

In keeping with his ethical thought, Kant proclaimed the supreme principle of law should be as follows:

> Every action is right and just, the maxim of which allows the agents free-
> dom of choice to harmonize with the freedom of every other, according to
> a universal law.[70]

We can immediately appreciate that the preservation of individual choice—
which also allows the prospect of social harmony—was central to Kant's
conception of justice. He moved from this principle very quickly to a formula resembling the law of equal freedom, variants of which are later to
be found in Herbert Spencer, John Stuart Mill, John Rawls and others.
As Kant wrote:

> So act that the use of thy freedom may not circumscribe the freedom of any
> other.[71]

The formula is not as capacious as Spencer's, which extended to each individual the *maximum* liberty compatible with a like liberty in others. But
we can discern here a sharp distinction, as is common in the classical liberal
tradition, between *liberty*, which is respectful of the same in others, and
license, which is not, a distinction drawn clearly in Locke as well.

Unfortunately, the precise boundaries between liberty and license cannot be surveyed infallibly, not given the relatively weak state of our knowledge today. In his essay, "Idea of a Universal History from a Cosmopolitan
Point of View," Kant wrote:

> Since Nature has set only a short period for [man's] life, she needs a perhaps
> unreckonable series of generations, each of which passes its own enlighten-
> ment to its successor in order finally to bring the seeds of enlightenment
> to that degree of development in our race which is completely suitable to
> Nature's purpose. This point of time must be, at least as an ideal, the goal
> of man's efforts.[72]

As with ethics, amelioration is a key goal of Kant's politics. But why do
any laws exist that are socially enacted and thus (obviously) exterior to the
individual will? If we wish a maximal, equal liberty, should we not abolish
all law? Kant's answer reveals much about his social theory in general, and
particularly that theory's evolutionary character:

> The very notion of law consists in that of the possibility of combining uni-
> versal mutual co-action with every person's freedom.[73]

Law exists to facilitate cooperation—but only on the condition that we do not at the same time obliterate anyone's freedom. Government does not exist to achieve a predetermined outcome in society, but to allow both voluntary cooperation and individual liberty, with which many different projects might be realized, and to contribute to the realization a future perfection that we living in the present cannot yet grasp.

In the same essay, Kant suggested that both freedom and cooperation were necessary for mankind *to achieve its destiny as a species*—an idea that we can forgive our readers for shrinking from, at least at first. It sounds nearly totalitarian, and almost like a sneaking return to certain of the comprehensive, hoped-for, and vain technologies of government that we have detailed. Yet on closer examination, the specific destiny that Kant imagined may not be so threatening after all. He wrote:

> The means which Nature employs to bring about the development of all the capacities implanted in men, is their mutual Antagonism in society, but only so far as this antagonism becomes at length the cause of an Order among them that is regulated by Law.[74]

It is irrelevant whether "Nature" causes this order to emerge, or whether the order emerges of its own accord, or even whether these two possibilities are just different ways of saying the same thing. What is key is that mankind can develop to the fullest only in the context of an ongoing social order, one that lasts across many lifetimes and permits a measure of peaceful competition, of "unsocial sociability," as Kant termed it.

One key locus of "unsocial sociability" is, of course, the market process, as even its defenders will admit. One cannot properly belong to the classical liberal tradition without a robust account of private property that entails its relatively unrestricted usage and transfer, and that thus entails the market process, about which we have said little so far. Our silence, though, is for a fairly strong reason: Kant himself would appear to have cared little for economics. He rarely deployed examples that proceeded from what we might term economic behavior in the narrow sense, that is, those involving buying and selling. His thoughts on the matter would appear to have been few, and we are left to draw inferences from a meager data set.

When Kant did write about market exchange, however, he certainly did not write to condemn the practice in all cases. Instead, he condemned only specific types of exchange, including instances of fraud, price discrimination, and the sale of organs, such as teeth, a common practice at the time. (Of these three, it is not at all clear that the last two condemnations must stand.) Kant was also aware of the writings of Jean-Jacques Rousseau, who did condemn commerce, often quite explicitly, and it is evident that while Kant had every opportunity, he declined to agree.

In all, however, much work remains to be done in theorizing the market process from the standpoint of Kantian ethics. Important groundwork has been laid by the contemporary philosopher Mark D. White, who suggests that Kant's *negative duties*—those things that we are absolutely obliged to refrain from doing to others—and Kant's *positive duties*—those things that we are obliged, in a limited sense, to do out of benevolence to others—are both compatible with a market society. The former make a market society possible, and the latter make it agreeable, as White has argued. White has even loosely paralleled these two types of duty with the worlds described in Adam Smith's two great works, *The Wealth of Nations* and *The Theory of Moral Sentiments*: The former considers the realm of merely negative duty, while the latter examines our positive duty of beneficence. Although Smith's ethics were not deontological like Kant's, the approach still seems promising.[75]

Kant's theory of the origin of property was in many ways more historically grounded than Locke's, or even Hume's. It also sat well with the intellectual and ethical project that we have outlined above—the gradual apprehension of the ethical laws of reason, and the reconciliation of the will to reason's dictates. Kant believed that property rights arose gradually, out of repeated claims, counter-claims, adjudications, and re-affirmations. Property did *not* arise from any one definitive act of claim, settlement, or assignment, whether by the state or by individuals.[76] This process-oriented view helps us gain new perspective on several vexing problems in property rights theory, including that of compensation for historical wrongs: It may prove, for example, that, contrary to the common law maxim concerning improperly acquired property, legitimacy *can* arise over time. Kant at least appears to claim as much. Given the initially arbitrary (and often criminal) origin of nearly all title to land, there is no other way forward in any case: As we discussed when considering Locke, it seems that we must either concede that all the world is stolen, after which we must establish an institution to redistribute everything, or we must admit that past errors are better corrected gradually. Institutions capable of redistributing everything may well be too dangerous in practice ever to be trusted, and thus our choice becomes clear. Gradual reform it shall be.

From an unowned condition, land in Kant's theory might first be appropriated by anyone with the means to defend it. No mixing of labor was required to stake a claim. This claim, however, in no sense conferred an absolute right. In this, Kant differed from Locke in two ways. Locke, recall, would insist that the mixing of labor was necessary to establish ownership. And from that point on, Locke held that ownership was settled and absolute.

On neither point would real life appear to correspond well to the Lockean account. On both, the Kantian account seems to describe initial appropriation with relatively greater historical accuracy. There is at best a *limited* obligation not to interfere with land that is in this manner only provisionally controlled by others; with the progress of civil society, land claims grow more settled and more binding. But the obligation to respect the claims of others may be breached if a land claimant refuses, for example, to enter into a state of civil society with his neighbors. It will not do to have barbarians on our borders. Indeed, this very consideration brought Kant to believe that implementing some social contract—any social contract—was morally obligatory, the contract's relative justness notwithstanding. Anything at all would constitute an improvement over the lack of civil society.

It is unclear to me how foundational this move is to Kant's social thought. After all, contracts of this type may be exceptionally rare.[77] In any case, though, the provisional obligation to respect property rights solidifies with the entry into civil society, which rational beings should recognize as desirable. It solidifies further with time and usage under a regime of sufficiently just laws.[78] As the modern scholar Marcus Verhaegh has written:

> [T]he best metaphor for Kant's account of movement out of the state of nature is one of disarmament—staged, negotiated disarmament. We are all duty-bound to reduce violent conflict and the potential for violent conflict by moving toward a scenario in which ownership disputes are decided by the rule of right law, rather than ongoing, competing military power. But prior to full disarmament—the fully cosmopolitan globe—military force plays a significant role in setting the bounds of right ownership.[79]

Alas. Military force is still necessary, and it would remain necessary, Kant believed, until a worldwide regime of perpetual peace had been established, one in which all countries enjoyed a republican form government, as well as the renunciation of war and standing armies. This cosmopolitan social order would be one of the crowning achievements of human civilization. Kant also thought it would take many generations to accomplish. In the meantime, governments should do their best to move toward it.

Kant's vision of a gradual transition—from violent appropriation defended violently, toward a cosmopolitan civil society defended by reason alone—anticipates much of subsequent German liberalism, and particularly the anarcho-capitalist vision of the early twentieth-century sociologist Franz Oppenheimer, whom we will examine below.[80] Yet while Kant made strong claims about the interior necessities of reason, he was relatively modest in his claims about the nature of history and its unfolding. He did *not* claim to have discerned a set of historical laws that will operate as of necessity, even if, at first glance, he may appear to have done so, and thus to stand condemned as a historicist. Unlike Marx or Hegel, Kant left room for, and indeed assigned a central place to, the liberty of individual action, which may, if allowed to operate, *eventually* instantiate the cosmopolitan society of the future. As we shall see, both Marx and Hegel proposed to detail the mechanisms by which the society of the future would come into being, but their mechanisms were ultimately not Kant's: Where Kant believed that a group of reasoning people living in a civil society across many generations constituted the discovery mechanism for eventual social perfection, Marx held that the discovery mechanism in question was to be found in impersonal economic forces, and Hegel proposed to locate it in the violent mutual conflict of nation-states. For Kant, though, the active agents are free individuals, not social classes, social forces, national spirits, or states, and the claims that Kant made about the future were few and qualified. His purported intellectual successors, Hegel and Marx, did not share Kant's intellectual modesty.

How, though, do property rights and cosmopolitan law relate to Kant's ethical project? Neither will necessarily make us good people; nothing, after all, prevents a propertied individual from having a bad will. Nor are property rights even necessary for possessing inward ethical freedom, for one *always* has the capacity to will the good—or not—regardless of how unfree or poor one may be. A good person may live, and be good, under a bad government, or in destitution. The traits are in this sense quite independent.

But property rights in things alone, and never in persons, can help us to obey more perfectly the negative duties that are most clearly implied by the second formulation of the categorical imperative: By granting each person the capacity to acquire, modify, and alienate property, we also allow them to use property to their own ends; we also declare, as it were, our maxim that only unreasoning things are to be used as tools, and never people. A cosmopolitan regime of private property that excludes slavery thus draws a bright line between the kingdom of ends, which is reserved for people, and the kingdom of means, which largely overlaps the legal category of property. This outward conformity, Kant believed, could lead people to the inner apprehension of the moral law, and at any rate, it made compliance less difficult for those of good will.[81]

Under a cosmopolitan property-holding regime, we likewise obtain a type of outward autonomy for ourselves. Much of the cosmopolitan project seems aimed at making the *outside* more closely resemble the *inside*. It aims at expanding our freedom of action in the phenomenal world, that is, the world of exterior experiences, so that it more closely resembles the freedom of action in the interior world of the mind. What we do with our property, do note, may be good or bad, but we will at least have secured one of the foundations of leading a morally good external life, which is the capacity for self-rule. (For Kant, the growing capacity for an adult-like self-rule, a rule independent of the state, was also the essence of the Enlightenment.[82]) Under free institutions, obedience to the written law and obedience to the moral law may now begin to harmonize, even if, as Kant warned, our current property claims may not be fully settled or just. In time they can be, if only we continue to will it.

Thus, at least a rough, outward set of prohibitions on certain uses of other moral agents merely as tools amounts to something like a legitimate purpose for government, at least for the time being. A classical liberal might likewise say that the positive duties commanded by the categorical imperative—such as the duty to treat others as ends in themselves—are not capable of being furthered by legislation: If one treats another as an end merely because the civil law has commanded it, then one has certainly not become a more moral person. One's will has not been made good, for the law only compels outward compliance.

UTILITARIANISM: A MODERN SHORTCUT

As David Conway observed in his book *Classical Liberalism: The Unvanquished Ideal,* political theorists beginning with Plato have frequently claimed, in addition to whatever else they might have asserted, that *happiness* is the purpose of government.[83] Although Conway's observation is correct, I have so far been unable to make much use of it. The reason is simple: As Conway himself went on to observe, definitions of happiness have varied enormously from one thinker to another. In consequence, observing that they nearly all take happiness to be in some sense the purpose of government does not substantially advance our inquiry. We must then go on to ask about each philosopher's conception of happiness, and when we do, we will generally find the same answers that we have elaborated above: Happiness means the devotion of oneself or one's resources to the community. Happiness means everyone knowing and taking up their allotted place in a transcendent order. Happiness means reducing the number and scope of social conflicts. And so on.

While some of these are nowadays eccentric notions of happiness, they were not always so. And Conway adds to these social accounts of happiness many individual ones besides: accounts of happiness that concern family life, religion, economic security, and the like. Conway moves from this diversity in our notions of happiness to an argument that judgments about happiness, and about the methods to be used in its pursuit, are irreducibly individual in character. Each sane adult is to be considered the best judge of his own happiness, and he is not to be considered well-suited to judge anyone else's happiness. A weighty presumption must be overcome to declare otherwise. From here, it is only a short step to favoring a liberal social order, reached through a contract-like agreement, with strong protections for economic and personal liberties. Such liberties would allow us to pursue our divergent aims in relative peace, and, when possible, in voluntary coordination. Conway's argument is worthwhile, but it will not be further elaborated here.

One distinctly modern treatment of the question of happiness and its social implications comes from the school of thought known as utilitarianism. Although utilitarianism does not necessarily prescribe any particular purpose for the government to follow, it does propose a method by which government purposes may be identified and tested. That is, committing to utilitarianism does not commit one to directing the state toward specific policies. Instead, it commits one to asking whether particular policy choices conduce to greater overall happiness, and to following what data can be found in answering this question. Utilitarian approaches to public policy analysis enjoy today a position of overwhelming preeminence in the political discourses of the developed world. Utilitarianism therefore deserves some consideration in our study, even while a full treatment of the history of utilitarianism, or a close consideration of the various types of utilitarian ethics, cannot be given.

As opposed to many philosophies of happiness that ran before it, utilitarianism makes two distinct moves. First, it posits that an individual's happiness consists purely of the sum of his individual pleasures, minus (in some sense) the sum of his individual pains. Both pain and pleasure are to be suitably qualified, and debate about the weighting of various sorts of pleasure and pain is not only permitted, but encouraged, in the hopes of coming to a better final reckoning of where our happiness stands.

Second, utilitarianism proposes to equate the social good with the maximizing of the summed happiness of all people in a society considered equally. As utilitarianism's great exponent Jeremy Bentham (1748–1832) put it, "The interest of the community then is, what?—the sum of the interests of the several members who compose it."[84] Communal interest for Bentham can only arise from individuals' interests. The happiness of the community properly considered is therefore not to be derived from a transcendent order, or from a General Will as intuited by the virtuous, or from some purported inner logic of history. The desires of a community are *composite* and *present*; they also have a reasonable claim to being *objective*, in the sense that they require only research to discover, rather than metaphysics, virtue, or statesmanship; and in that each individual is well-positioned to examine at least one datum exhaustively.[85] Bentham frequently reiterated this definition of communal interest, and it has become a commonplace in modern public policy analysis either to presume that he was essentially correct or to adopt a closely parallel method—one that aims, for example, at maximizing the gross domestic product, which is certainly easier to measure.

Present-day utilitarianism also commonly extends this notion of summed collective good to all of humanity and even to all creatures capable of experiencing pleasure or pain. Such extensions are controversial, but they are in keeping with utilitarianism's stated concerns. Indeed, Bentham himself argued in favor of animal rights and speculated that the day would come when humanity would view animal cruelty as an evil to be suppressed.

Bentham took yet another highly significant step when he posited a one-to-one identity between those things that brought net happiness and those that were useful. This was an unusual step at the time. Formerly, those things that brought pleasure immediately were apt to be termed amusements, and these were explicitly excluded from the useful arts, whose purpose was, purportedly, not to amuse but to secure some higher form of value.

Bentham agreed that both should be considered good, but he insisted that they were both good for the same underlying reason: Some endeavors would take longer than others to achieve their reward, but when they did, in all cases the reward would be found to consist of pleasure. Consider this striking passage from Bentham's *The Rationale of Reward*:

> Custom has in a manner compelled us to make the distinction between the arts and sciences of amusement, and those of curiosity. It is not, however, proper to regard the former as destitute of utility: on the contrary, there is nothing, the utility of which is more incontestable. To what shall the character of utility be ascribed, if not to that which is a source of pleasure?[86]

By these terms, utility is to be found in anything that gives us pleasure, whether that pleasure comes sooner or later. As a result, we ought not to blush at calling amusements *useful*. They are undoubtedly so, and they lose nothing by yielding their pleasures the most immediately. Other things give their pleasures over a longer timeframe, but the reason for their utility is at root the same.

Ludwig von Mises once observed, "that ethics, no matter how strict an opponent of eudaemonism it may at first appear to be, must somehow clandestinely smuggle the idea of happiness into its system."[87] All ethical systems seemingly profess to yield happiness in some sense or another; at the very least, they ask adherents to be happy about instantiating the system. In this, utilitarianism would appear to distinguish itself only in that it takes happiness to mean summed pleasures minus summed pains, and that an account of each may be given at nearly any point in an individual's life. It may also pride itself on being more honest than the others about its ultimate aims.

One may fairly wonder, however, whether this approach robs the word happiness of all meaning, in that it makes the search for happiness mean something entirely too close to a generic term for motivation. And thus, we may have a tautology: In the end, we are all made happy by whatever it is that makes us happy, which tells us nothing. And saying that we should maximize it seems to add very little. *Of course* we are motivated by that which motivates us.

Thinkers of previous eras certainly did not make this move. Recall how Herodotus made Solon describe the happiness of the Athenian Tellus; the story of Tellus would seem to defy utilitarian analysis, although it would also seem unfair to dismiss it as an account of happiness. It is simply a conception of happiness that fits badly with utilitarianism: In the narrative, Tellus could not be called happy, or unhappy, until after he was dead. Upon which he experienced presumably neither pleasure nor pain. With evident approval, Herodotus has Solon declare that no one's happiness is known for certain until death, not even if one is literally as rich as Croesus. Herodotus tells us that Croesus could *not* be called happy, insofar as he did not die happy, which is all that matters.

By this conception of happiness, which is *not* built of increments of pleasure and pain, the utilitarian analysis of public policy, or even of individual action, is impossible: It can't be seriously asserted that everyone should immediately rush off to die in war so as to win happiness. The implication emerges that the thing Herodotus described as happiness would necessarily arrive only rarely, and that it would never be knowable to its possessors in this lifetime. It is moreover absurd to attempt to describe death in battle as *pleasurable*—and yet pleasure is for utilitarians the constituent stuff of happiness. Death in battle may be honorable perhaps, or brave, or even a source of happiness when we contemplate the prospect in advance of its arrival. But death in battle is not pleasurable. Indeed, if death in battle were merely pleasurable, and if it entailed no prospect of pain or loss, we might be completely unable to derive the same sort of happiness from it that Tellus did, because it could not serve as a test of our honor or bravery, which it had been for him.

The idea that pleasures should be represented as credits, and pains as debits, and that a running summation should be made of them, and that this summation should be identified with happiness, was likewise alien to the Epicureans and the Stoics, who both held that happiness properly understood did not derive from anything outside of one's own mind and character. Get these right, they held, and happiness would follow. Get them wrong, and there would be suffering.[88] In neither case did the community, or particularly the government, or even one's circumstances in most other regards have much to do with it. Christians, meanwhile, made room in their ethics for nuance about both worldly pain and worldly pleasure. Neither was always embraced or always rejected; and although Christians seek a goal that might be endorsed in utilitarian terms—provided we admit the existence of Heaven—the idea of seeking Heaven merely for its pleasures has always had an evil reputation in Christian thought.

Utilitarianism took part in two further modern intellectual developments. First, it was exemplary of the economizing turn in political thought, by which the tools and methods of economics were increas-

ingly brought to bear on, and considered germane to, the art of governing. Whereas formerly governments might be described as having a form
similar to a family, or as being like a human body—or as resembling the
heavens themselves—now they were, and are, increasingly likened to a
business, with questions of budget, finance, and demography coming to
the fore. The burgeoning discipline of economics supplied the methods
needed for this new approach to government, while utilitarianism supplied the moral framework, and it did so in terms that were highly compatible with this new, business-like approach. Indeed, the two sometimes
overlapped: Substitute wealth for pleasure, and Bentham's account of
the community's good is identical to his account, given elsewhere, of
a community's wealth. The latter is no more than the summed total of
individual wealth, and it is obtained by the reckoning of individual credits
and debits.[89]

Second, for all its interpersonal summations, utilitarianism is still in
some key respects an individualist and egalitarian approach: It is, after all,
individual happiness that we are asked to sum, and each individual should
in theory receive his or her equal due in all social calculations. The social
good is reduced to a composite; what is good for society cannot even be
meaningfully considered without direct reference to individuals, with each
individual considered to be the equal in worth of any other. As a result,
conflicts between any proposed policy furthering the social good and the
interests of individual members must be satisfyingly addressed before a
utilitarian can assent to the policy. Many putative social goods will immediately be ruled out owing to their conflict with the principle of securing
the greatest happiness for the greatest number. And all policies whose
rationales explicitly rank the happiness of one type of individual above
the happiness of some other type of individual must either seek out new
rationales or else be abandoned.

The consequences in the real world were enormous. Utilitarians,
Bentham included, were at the forefront of equal legal treatment for all
people, regardless of gender or racial difference, and it is easy to understand why: If, for example, the laws of marriage render men happy but
women at least proportionately unhappy, the results can't be called anything better than indifferent. They are in no sense good. Because happiness can be attained (or not) by anyone, utilitarian thought has tended
strongly to encourage egalitarianism: The happiness of one is not to be
favored over the happiness of another merely on account of some other,
unrelated trait.

Bentham himself favored free trade, free expression, abolition of slavery, abolition of penalties for usury, and the separation of church and state, and he justified each with reference to his principle of the greatest good for the greatest number. In short, he used the methods of utilitarianism to argue for what amounted to classical liberal ends.

Whether utilitarianism always yields classical liberalism is now an important question. When one adopts the utilitarian method of determining the purposes of government, the menu of possible aims for government must necessarily shrink, even if the resulting program is not always classical liberal in character. Indeed, the menu of possible aims may shrink all the way down to one single item: If government itself can provide the greatest happiness for the greatest number in some comprehensive sense, then it must do so—with the significant caveat that this may constitute no real-world shrinkage of the government at all. Meanwhile, it may even be shown that having a government in itself is contrary to the utilitarian imperative. If that's the case, then the government is morally obliged to abolish itself. Simply adopting a utilitarian framework does not predetermine any of these answers, however. Much data must be obtained before settling on any of these choices, and it is unclear how that data can be obtained.

Other problems soon arise as well. David Conway's argument, outlined above, suggests that we can never do utilitarian calculations as well as we might like in the real world, and that we will never be able to agree either on what constitutes happiness or on how we can best obtain it. Reasonable people will probably always have different answers to these questions, and no objective way of settling their disputes is likely to be found. They may always call one another unreasonable, but, in doing so, they risk returning to previous modes of thinking about happiness; that is, they risk abandoning several of utilitarianism's most attractive features, including its commitment to interpersonal egalitarianism and its promise to elucidate social interests by basing them on the summed interests of individuals.

Cynically, Herbert Spencer once suggested that the only method of resolving these disputes about the true meaning of happiness would have to be state intervention—a solution that would leave virtually no one happy:

> If each man carried out, independently of a state power, his own notions of what would best secure "the greatest happiness of the greatest number," society would quickly lapse into confusion. Clearly, therefore, a morality

established upon a maxim of which the practical interpretation is question-
able, involves the existence of some authority whose decision respecting it
shall be final—that is, a legislature.

Not, mind you, that the state would do a good job of it. But at least
state power would prevent any actions to the contrary.[90] Only through
such heavy-handed means, he thought, could utilitarianism be put into
practice—yet if it were enacted in this manner, utilitarianism would be
a sham. As a result, Spencer ultimately settled on something very close
to Conway's later agnosticism about precisely what would render various
individuals happy. Spencer likewise settled on classical liberalism.

A general tendency can be observed in much utilitarian public policy
analysis: Insofar as the investigators are agnostic about the appropriate
means of securing greater aggregate happiness, or about what constitutes
happiness for individuals, they will tend toward solutions that favor individual self-determination within a framework of private property and formal legal equality, just as Conway and Spencer have done. Insofar as the
investigators believe that they have already determined the most appropriate means of attaining greater aggregate happiness, they will abandon
these classical liberal commitments in favor of providing the appropriate
means directly, through state action. Humble utilitarians tend to advocate
liberty. Confident utilitarians are a different species entirely.

And yet it is not necessarily clear, even in the presence of broad agreement about certain means of securing greater aggregate happiness, that
the optimal course is necessarily direct state action to supply the means
in question. To say so is to make an additional inference, namely that the
state is the most efficient provider of these means. This chain of inference
is subject to falsification in at least two different ways: First, we may be
wrong about the content of the inference itself; that is, some people may
not actually be made happy by our government's proposed actions, and
their potential unhappiness must figure into our calculus. And second,
even supposing that a particular means-ends relationship is appropriate for
a sufficiently large number of people (however large *that* needs to be), the
state is not necessarily the most efficient means of attaining it.

Let us consider what would seem a highly unproblematic means-ends
relationship: Supplies of adequate food are necessary (a means) for the
happiness (the end) of all human beings. Apart from a few ascetics and
devotees of extreme self-sufficiency, there will be virtually no disagreement. But does it then follow that government must be the means of

supply? It has in some times and places been tempting to answer yes, and to conclude that one of government's purposes must be to set up a free public food distribution system.

But this neglects the crucial question of abundant food *considered as an end*, toward which various means may be directed, with greater or lesser efficiency. Nor will we necessarily know the complete list of possible means toward our intermediate end. Both of these considerations should incline us away from implementing a state-based solution, even while we need not abandon our normative commitment to utilitarianism: The state may not be the best method at hand; moreover, turning to it will tend to inhibit the possible discovery of other means of provision yet unknown to us.

For most Americans in the early twenty-first century, the question of free public food distribution isn't terribly difficult. We have come to associate state distribution of food quite closely with the few remaining instances of famine in the modern world—instances for which state distribution is directly to blame. Government would appear to be a relatively poor means to the delivery of this intermediate end. Identifying a widely shared agreement on a means-ends relationship is only one step among many in a well-considered utilitarian analysis of a given public policy.[91] When we consider the many chains of indeterminate means-ends relationships necessitated even by a fairly obvious means that conduces quite directly to happiness (such as food), our doubts about more contested ends can only multiply.

Within the classical liberal tradition, there has been much disagreement about the usefulness of utilitarianism. F.A. Hayek's argument against utilitarianism went well beyond Conway's or Spencer's. His critique is important for our inquiry, but explaining it requires a bit more background. The utilitarian approach in ethics had long been divided into two somewhat rival theories: *Act utilitarian* accounts stress that each individual action must be undertaken with a view to the greatest happiness for the greatest number. *Rule utilitarian* accounts seek to enunciate general rules whose obedience would attain the same end.

Between the two, act utilitarianism proposes to adhere more strictly to the original system as described by Bentham. Rule utilitarianism proposes to be easier to understand and teach, even if it sometimes risks some loss of utility in circumstances where following an otherwise good rule results in occasional suffering. In cases like these, rule utilitarians can at least console themselves that the rule is good in general; and that adverse circumstances are by definition rare; and that, as they believe, the teaching of rules that conduce to happiness is apt to bring the greatest happiness on the whole. Act utilitarians, meanwhile, would sooner break a rule rather than do anything that resulted in suffering. And their principle, as it were, does not seem terribly difficult to teach.

Hayek answered each prong in turn. To the act utilitarians, he argued as follows: Although act utilitarianism was the only version of the theory "consistent in basing ... approval or disapproval of actions exclusively on their foreseen effects of 'utility,' ... at the same time, in order to do so, it must proceed on a factual assumption of omniscience which is never satisfied in real life."[92] The full effects of our actions are almost never known, even as regards their merely physical outcomes, let alone the feelings they engender in others. Act utilitarians will therefore often find themselves in situations where they must affirm that they have no reason for preferring one action over another.

To rule utilitarians, Hayek made the following reply: "To judge the utility of any one rule would ... always presuppose that some other rules were taken as given and generally observed ... so that among the determinants of the utility of any one rule there would always be other rules which could not be justified by their utility."[93] Rules exist in systems; they depend on one another for their intelligibility and coherence, and much of what we judge as either good or bad about a given rule is only good or bad depending on the other rules in place at the time. It is not altogether clear that improving the utility of one rule at a time in sequence, while holding each other rule constant, will ever result in an optimum rule set. And it is for that matter unclear who will order the decision making that this process would necessarily entail. Perhaps, with Spencer, we may find the need for the state creeping back in to utilitarianism.

To Hayek, rule utilitarianism would seem to have developed without any consideration of why humans follow rules in the first place. The reason we follow rules, Hayek believed, was "our necessary ignorance of most ... particular facts." We have rules because we do not know, and cannot know, the full effects of our actions. We follow rules "irrespective of the known effects of the particular action" because doing so yields at least two important benefits to the equally ignorant agents around us.[94] First, social rules that have developed gradually over time are likely to have become well adapted to the society in which they are found; they represent at the very least the educated guesses of many generations, from which many atrocious or incompatible rules have already been weeded out, in a manner quite like the one found in Gratian's *Decretals*. Second, social rules are to be followed because they help other individuals form accurate expectations about our own actions. Social rules therefore coordinate human actions, not with any one grand outcome in mind but with a view to coordinating and harmonizing multiple lesser outcomes, which may be individually chosen.

As Hayek wrote, "[Ethical rules] serve not to make any particular plan of action successful, but to reconcile many different plans of actions."[95] Utilitarianism constituted for him an abandonment of this coordinating function of ethical rules. Rules adopted under utilitarianism would probably not aim at coordination, but rather at achieving a particular and comprehensive plan. Probably the plan of a legislator, as Spencer had warned.

As a result, Hayek rejected utilitarianism. "Utilitarianism, to succeed in its aims, would have to attempt a sort of reductionism which traces all rules to the deliberate choice of means for known ends ... But in fact nobody has yet achieved such a full understanding or succeeded in reconstructing an altogether new system of moral or legal rules from the knowledge of the needs and the effects of known means."[96]

Following his lead, we may say that at least some of the stronger forms of utilitarianism come close to proposing a totalizing science of government, one that, like others before it, entails making a set of knowledge claims that are ultimately untenable. Now, utilitarianism is *not* astrology. It is a respectable and sophisticated attempt to solve very difficult interpersonal ethical problems. Like all such attempts, it is contested and in some ways flawed. But it is not flim-flam. It is an invaluable tool in the ethicist's toolkit. Still, though, it is apt to be abused in particular ways, and these might make it into something that it ought not to be.

Such misuse may not be the fault of utilitarianism in the abstract. It may be the fault of human diversity, which utilitarianism insufficiently considers. We humans may just be too different from one another, and the aims we pursue may just be too discordant, to formulate any one plan that could keep us all happy. Such considerations may lead one to an opinion similar to Hayek's. But there is another answer: Might we eliminate the diversity? If it conduced to happiness, might that be a small price to pay?

It is surely true, as Bentham observed, that children are more easily amused than adults. But if this is so, and if the good consists of securing pleasure—then why not resolve to make our adults more childlike, the better to make them happy? It would certainly make the utilitarian ruler's task a great deal easier, and nothing within utilitarianism would appear to prohibit it. Yet the resulting society is very difficult to endorse.

Several portraits of this type of society exist, including H.G. Wells' *The Time Machine*, but the greatest of them is Aldous Huxley's *Brave New World*. The inhabitants of Huxley's dystopia are repeatedly described as childlike, their grossly sexualized amusements notwithstanding. Even that society's alphas—the highest caste, selected eugenically and conditioned to superior intelligence—were taught to remain childlike whenever possible. Tragically, some of them were aware of their condition. As one disgruntled alpha put it, they were to be "[a]dults intellectually and during working hours. Infants where feeling and desire are concerned."[97] Childlike amusements did not amuse him, however, and his story ends tragically.

One of the most trenchant critiques of the utilitarian project, then, is that only certain kinds of pleasure can be mass produced. These are inevitably among the shallowest, and they are unlikely to be appreciated unstintingly by those of an intellectual temperament. To pursue the utilitarian project may require, not the extinguishing of all intelligence, necessarily, but the extinguishing of a particular temperament that often accompanies intelligence, and that we should not wish to be without: a critical, reflective, doubting, and self-consciously dissatisfied stance toward the world. It is not so easy to subsume this temperament under the rubric of pleasure-seeking, whether of an immediate or a deferred kind, not without making what will likely appear an ad hoc pronouncement—that a self-consciously dissatisfied temperament *really is pleasurable*, despite all appearances to the contrary, and that we are happy when we are sincerely dissatisfied about a deplorable society.

By contrast, a purposefully dumbed-down and childlike society might fit the utilitarian bill all too well, if only sufficiently uncritical inhabitants can be found to populate it. In David Conway's terms, there would no longer be any diversity of understanding about what renders us happy, and this—though at a terrible cost—could make utilitarianism work, at least in a narrow sense. If anyone were to protest alone, or in the company of a tiny minority of dissatisfied intellectuals, they would not be very convincing. Their unhappiness could be set aside, perhaps, with an extra dose of *soma*.

NOTES

1. Harro Höpfl and Martyn P. Thompson, "The History of Contract as a Motif in Political Thought," p. 920. http://www.jstor.org/stable/1904609.
2. Skinner, *The Foundations of Modern Political Thought*, vol I p. 159.

3. For a dissenting analysis, one that emphasizes the importance of divine agency in Suárez's thought, see Sommerville, "From Suarez to Filmer."
4. Berman, *Law and Revolution*, p. 393.
5. Skinner, *Hobbes and Republican Liberty*, p. 112.
6. John Milton, *The Tenure of Kings and Magistrates*, https://www.dartmouth.edu/~milton/reading_room/tenure/title.shtml.
7. Hobbes, *English Works*, http://oll.libertyfund.org/titles/585#Hobbes_0051-03_621.
8. Ibid., http://oll.libertyfund.org/titles/585#Hobbes_0051-03_724.
9. Ibid., http://oll.libertyfund.org/titles/585#Hobbes_0051-03_300.
10. Ibid., http://oll.libertyfund.org/titles/585#Hobbes_0051-03_411.
11. Ibid., http://oll.libertyfund.org/titles/585#Hobbes_0051-03_490.
12. Leo Strauss, "Notes on *The Concept of the Political*," *in* Carl Schmitt, *The Concept of the Political*, pp. 106–107.
13. See Heyd, "Hobbes on Capital Punishment."
14. Hobbes, *English Works*. http://oll.libertyfund.org/titles/585#lf0051-03_label_089.
15. Ibid., http://oll.libertyfund.org/titles/585#lf0051-03_label_069.
16. Ibid., http://oll.libertyfund.org/titles/585#lf0051-03_label_089.
17. Skinner, *Hobbes and Republican Liberty*, p. 167.
18. Hobbes, *English Works*. http://oll.libertyfund.org/titles/585#Hobbes_0051-03_487.
19. Hobbes supplied many idiosyncratic definitions to common terms, but he did not define the word "danger" anywhere in *Leviathan*. His definition of "honour," however, is puzzling when we recall it in this context. In chapter 10, he wrote: "To value a man at a high rate, is to *honour* him; at a low rate, is to *dishonour* him. But high, and low, in this case, is to be understood by comparison to the rate that each man setteth on himself." Conceivably, a population might render more and more acts of government illegitimate simply by adopting a higher opinion of itself, which would render more state acts dishonorable. See Hobbes, *English Works*, http://oll.libertyfund.org/titles/585#Hobbes_0051-03_196.
20. Ibid., http://oll.libertyfund.org/titles/585#Hobbes_0051-03_481.
21. Ibid., http://oll.libertyfund.org/titles/585#Hobbes_0051-03_1264.
22. Ibid., http://oll.libertyfund.org/titles/585#Hobbes_0051-03_474.
23. Ibid., http://oll.libertyfund.org/titles/585#Hobbes_0051-03_433.
24. Skinner, *Hobbes and Republican Liberty*, p. 151.

25. Hobbes, *The Elements of Law Natural and Politic*. ch 23 section 9. http://oregonstate.edu/instruct/phl302/texts/hobbes/elelaw. html.
26. For a further defense of this claim, which does require some close reading, see Skinner, *Hobbes and Republican Liberty*, pp. 75–81.
27. Skinner, *Hobbes and Republican Liberty*, p. 163. Emphasis in original.
28. Laslett in Locke, *Two Treatises*, p. 6.
29. Ashcraft, *Revolutionary Politics*.
30. Philip Milton, "John Locke and the Rye House Plot," p. 660.
31. J. R. Milton, "Dating Locke's Second Treatise."
32. As Wootton summarizes: "[T]he evidence suggests that the *First Treatise* was, as Laslett, [Richard] Ashcraft, and [John] Marshall agree, an Exclusion tract, and was largely written by March 1681 … But the *Second* came second, as Ashcraft and Marshall recognize." Wootton, ed. *John Locke: Political Writings*, p. 63. The nature of Laslett's error is discussed in Menake, pp. 609–612.
33. Locke, *Second Treatise*, ch 2.
34. Locke, *Second Treatise*, ch 2 s. 6.
35. Locke, *Second Treatise*, ch 2 ss 7–8.
36. Locke, *Second Treatise* ch 2 s 13.
37. Laslet in Locke, *Two Treatises*, pp. 102.
38. Ibid.
39. Laslett in Locke, *Two Treatises*, p. 103.
40. Locke, *Second Treatise*, ch 9 s 123.
41. Locke, *Second Treatise* ch 5 s. 37.
42. Rothbard, *Ethics of Liberty*, p. 58.
43. Hume, "Of the Original Contract," in *Essays*, p. 480.
44. See Schaefer and Schaefer, "An Innovative Approach to Land Registration." A thought experiment also helps: If we imagine the absence of government, we find ourselves immediately thinking neither that we will have forfeited our property, nor that we must in justice relinquish it to some other individual or group, but rather that we must find some new (and perhaps less satisfying) way of securing it.
45. David Friedman, *The Machinery of Freedom*.
46. Rousseau, "Constitutional Project for Corsica," in which he writes, "The abuse of political institutions follows so closely upon their establishment that it is hardly worth while to set them up, only to

see them degenerate so rapidly." http://www.constitution.org/jjr/corsica.htm.

47. Rousseau did draft two constitutions, never implemented, for Poland and Corsica. Rousseau scholar Judith Shklar characterizes the Polish project as a "superficial fantasy; as Rousseau well knew." Shklar, *Men and Citizens*, p. 176.

48. The definitive modern discussion of Rousseau's social thought is in my view Shklar, *Men and Citizens*. Readers should consult this work for much more detail on Rousseau's social thought, particularly on Rousseau's ideas about religion, family, childrearing and education, which cannot be discussed in detail here.

49. Rousseau, "A Dissertation on the Origin and Foundation of the Inequality of Mankind," in *The Social Contract and Discourses*, http://oll.libertyfund.org/titles/638#Rousseau_0132_654.

50. The gendered aspects of Rousseau's thought are deeply significant, and my choice of the word "effeminate" is not arbitrary. Rousseau belonged to an intellectual tradition, classical republicanism, that denounced effeminacy in men as one of the leading symptoms of social decay; in this, he owed much to Machiavelli. See Schwartz, *The Sexual Politics of Jean-Jacques Rousseau*, pp. 46–47 and throughout.

51. Rousseau, "A Dissertation on the Origin and Foundation of the Inequality of Mankind," in *The Social Contract and Discourses*, http://oll.libertyfund.org/titles/638#Rousseau_0132_725.

52. Rousseau's influence on ethnology has been vast, as anthropologists are quick to acknowledge. "Not only did Rousseau predict ethnology, he founded it," wrote no less a personage than Claude Lévi-Strauss in his memorial lecture "Jean-Jacques Rousseau, Fondateur des sciences de l'homme," available at http://www.espace-rousseau.ch/f/textes/levi-strauss1962.pdf. Lévi-Strauss deemed Rousseau's wildly unempirical *Second Discourse* "the first general treatise of ethnology," a bit of praise that may seem quite unfairly to indict the entire discipline: genuine ethnology requires months and years of difficult observational work.

53. Shklar, 42.

54. McAlpin, "Innocence of Experience."

55. Rousseau, "A Dissertation on the Origin and Foundation of the Inequality of Mankind," in *The Social Contract and Discourses* http://oll.libertyfund.org/titles/638#Rousseau_0132_718.

56. Shklar, p. 104.
57. Jean-Jacques Rousseau, *Confessions*, book x, trans. by the author. French text available at http://athena.unige.ch/athena/rousseau/confessions/rousseau_confessions_10.html.
58. Shklar, p. 52.
59. Rousseau, "A Dissertation on the Origin and Foundation of the Inequality of Mankind," in *The Social Contract and Discourses*. http://oll.libertyfund.org/titles/638#Rousseau_0132_684.
60. Ibid.,http://oll.libertyfund.org/titles/638#Rousseau_0132_738.
61. Ibid.,http://oll.libertyfund.org/titles/638#Rousseau_0132_740.
62. Ibid., *The Social Contract*, http://oll.libertyfund.org/titles/638#Rousseau_0132_121.
63. Cohen. *Rousseau: A Free Community of Equals*, p. 16.
64. Ibid., p. 66.
65. Rousseau, *Social Contract* Book IV Ch 2, http://oll.libertyfund.org/titles/rousseau-the-social-contract-and-discourses#Rousseau_0132_740.
66. On this point, see Darnton, "Readers Respond to Rousseau," pp. 215–256, but especially 231 and onward.
67. Modern classical liberals have had an uneasy relationship to Kant, though it is not entirely clear why this should be. For example, it may be that Ayn Rand's quarrel with Kant arose not from his ethics or politics, but from his metaphysics and epistemology. This, while plausible, would sit badly with the few relatively clear places in which Rand attempted to critique Kant; in them, she makes primarily ethical objections. Metaphysical and epistemological concerns, meanwhile, are beyond the scope of our inquiry.
68. Kant, *Fundamental Principles of the Metaphysic of Morals*, p. 29, in *Kant's Critique of Practical Reason and Other Works on the Theory of Ethics*. Note that the title of the *Groundwork* is here translated as "*Fundamental Principles...*"
69. Nozick, pp. 33–34.
70. Kant, *Metaphysics of Ethics* http://oll.libertyfund.org/titles/1443#Kant_0332_338.
71. Kant, *Metaphysics of Ethics* http://oll.libertyfund.org/titles/1443#Kant_0332_341.
72. Kant, "Idea for a Universal History from a Cosmopolitan Point of View," second proposition. In *Kant's Principles of Politics*, http://oll.libertyfund.org/titles/358#Kant_0056_36.

73. Ibid.
74. Kant, "Idea for a Universal History from a Cosmopolitan Point of View," fourth proposition. In *Kant's Principles of Politics,* http://oll.libertyfund.org/titles/358#Kant_0056_39.
75. See Mark D. White, "Adam Smith and Immanuel Kant."
76. This summary follows closely Marcus Verhaegh, "Kant and Property Rights."
77. Dodson, "Autonomy and Authority in Kant's Rechtslehre."
78. Verhaegh, 20.
79. Ibid, 21.
80. Oppenheimer, *The State.*
81. That is, Kant did *not* really think that external conformity to the moral law was valueless. Although it could not be called ethically foundational, external conformity did have a didactic and instrumental value. See Kant, *Critique of Practical Reason,* Part Second, "Methodology of Pure Practical Reason," pp. 209–217 in *Critique.*
82. Kant, "What Is Enlightenment?"
83. Conway, *Classical Liberalism.*
84. Jeremy Bentham, *An Introduction to the Principles of Morals and Legislation* (Oxford: Clarendon Press, 1907). http://oll.libertyfund.org/titles/278#bentham_0175_45.
85. Their discovery, of course, is full of difficulties all the same, some of which we will discuss below. In many instances, markets may act as mechanisms for the discovery of this type of communal will, but, as Hayek notes, this discovery will always come in the context of a given order composed of social rules and institutions. The type of communal interest posited by utilitarianism may therefore be less objective than it appears.
86. Jeremy Bentham, *The Rationale of Reward* in *Works,* vol 2. http://oll.libertyfund.org/titles/1921#Bentham_0872-02_3579.
87. Ludwig von Mises, *Epistemological Problems of Economics,* p. 138.
88. Consider this from the *Sovran Maxims* of Epicurus: "It is impossible to live a pleasant life without living wisely and well and justly, and it is impossible to live wisely and well and justly without living pleasantly. Whenever any one of these is lacking, when, for instance, the man is not able to live wisely, though he lives well and justly, it is impossible for him to live a pleasant life." In Diogenes Laertius, *Lives of Eminent Philosophers,* "Epicurus."

89. Jeremy Bentham, *Works*, http://oll.libertyfund.org/titles/1922#Bentham_0872-03_284. Regard for the wealth of individuals comprising the community was a fraught subject in utilitarianism; policies that foster the private accumulation of wealth are one thing, while policies that treat all wealth as national—and thus as properly subject to government control, as well as maximization—are another.

90. Herbert Spencer, *Social Statics*, http://oll.libertyfund.org/titles/273#Spencer_0331_46.

91. The act of work itself can be a source of happiness to the worker, not merely for the money it provides but also in the construction of a worthwhile life. Permitting the search for diverse and meaningful forms of work may therefore be necessary to a society's pursuit of happiness. See Tomasi, *Free Market Fairness*, pp. 89–91.

92. Hayek, *Law, Legislation, and Liberty*, vol II, p. 20.

93. Ibid.

94. Ibid., p. 21.

95. Ibid., p. 22.

96. Ibid., pp. 22–23.

97. Huxley, *Brave New World*, p. 93.

CHAPTER 7

The Modern Omnipotent State

HEGEL AND THE WORLD SPIRIT

G.W.F. Hegel wrote famously daunting prose, but the ideas he advocated are often quite familiar. Once one knows what Hegelian thinking looks like, one finds bits and pieces of it everywhere. In part this owes to Hegel's enormous influence; in part it owes to the fact that many aspects of his thought were by no means original. If students are encouraged to read Hegel with a patient guide, they will often find in him a confirmation of many of their prior beliefs.

Such are the risks we run. Hegel represents nearly everything that classical liberals and libertarians have always fought against, and I believe they have ultimately been right to do so. Ayn Rand thought Immanuel Kant was the most dangerous philosopher of all time, but I personally would choose Hegel. He stands nearly at the origin of the modern cult of the state, and he has had a pervasive influence on both the left and the right of modern politics.

Let's start at the beginning, then, with the foundational concepts.

Hegel believed that history had a purpose, or a meaning, and that he had grasped this meaning, at least insofar as it was possible for any one person of his time or place. History for Hegel proceeded as a series of conflicts and resolutions. Inwardly, on the intellectual plane, the progress of history was a march through a series of progressively more refined truths, with each new school of thought or philosophical synthesis building on

© The Author(s) 2017
J. Kuznicki, *Technology and the End of Authority*,
DOI 10.1007/978-3-319-48692-5_7

the ones before it, subsuming them, and integrating them into an ever-greater whole, in a progressive approach toward *absolute* truth.

Truth, for Hegel, was not a set of static conclusions or of timeless, changeless ideas. Truth was a lived experience, together with the awareness that can only come from having gone through that experience. Truth was therefore inseparable from *history*, and trying to apprehend truth without its history would be like trying to become an author without writing anything. One might imitate authors as much as one liked. One might say the things that authors say, or hang out in all the same places that they do. But this would not make one an author. Truth, which Hegel also called the Absolute, had to be worked out through practice. There was no other way.

None of this is terribly unreasonable. Much human knowledge is after all tacit; in many and perhaps most fields there simply is no substitute for lived experience. But Hegel appears to have aimed at something more than just tacit knowledge. At times his attempts to describe it verge on the mystical. Here, for example, is an attempt that he made in the preface to his most important work, *The Phenomenology of Spirit*:

> The True is the whole. But the whole is nothing other than the essence consummating itself through its development. Of the Absolute it must be said that it is essentially a result, that only in the end is it what it truly is; and that precisely in this consists its nature, viz. to be actual, subject, the spontaneous becoming of itself ... the words 'the Divine,' 'the Absolute,' 'the Eternal,' etc., do not express what is contained in them; and only such words, in fact, do express the intuition as something immediate. Whatever is more than such a word, even the transition to a mere proposition, contains a becoming-other that has to be taken back, or is a mediation ... Reason is, therefore, misunderstood when reflection is excluded from the True, and is not grasped as a positive movement of the Absolute.[1]

Which is opaque to say the least. One attempt at stating it in plain English is as follows: Whenever we consider a truth, we must remember that the True is the whole—which means that the True includes the historical process producing it. The True also includes our process of understanding it, which begins, but does not end, in mere words and propositions. Any important truth's process of development is replete with setbacks, confusions, and even contradictions, all of which we will necessarily encounter as well, when we learn about it. None of this can be collapsed into mere words. Only experience will do.

Hegel went further: To be worthy of the name, Truth also has to negate and subsume all the contradictions that arise in the process of experience. It has to vanquish those opposites and then take on their attributes, preserving them while destroying them at the same time. Hegel termed this process *sublation* (*Aufheben*). Through it, truth became multifaceted, and contradictions were resolved.

Although popularizers of Hegel have used the terms *thesis, antithesis,* and *synthesis* to describe the process of moving toward Truth through the setbacks and contradictions of experience, Hegel did not use these terms; he preferred to write of *ideal, negation,* and *actuality,* respectively, with actuality appearing only *after* the encounter with a conflicting principle. Because we must approach the Absolute through a long series of such threefold steps, the best we can generally hope for is to grasp a *moment* of truth, a technical term for Hegel and his followers that signified a limited aspect of truth. A moment is simply one facet of the development of a thing, the Absolute, which is multifaceted. Properly understood, the Absolute refers not only to a result or a proposition, but also to the entire process of arriving at it.

Unfortunately, any particular moment is liable to be mistaken for being the Absolute, whole and entire. This is why Hegel wrote at times of being *correct (richtig)* while not yet having attained Truth—correctness for him seems to have meant *accurately reflecting a moment.* But if one is wholly true to a moment, one cannot be wholly true to the Absolute.

Hegel believed that this dialectic, this process of truth through conflict, operated everywhere. He held that it worked inside the individual mind, in the physical sciences, and also in societies. In the latter, it gave rise to, and explained, the course of history.

Two famous examples of just how this dialectic works in practice can be drawn from Hegel's stylized history of the human spirit, presented also in the *Phenomenology.* The first example is what Hegel called the encounter with the Other; the second is what is sometimes termed the *master-slave dialectic.* Both bear directly on politics.

Hegel asserted that self-consciousness first arises from a process of mutual recognition that can take place only in the presence of an Other: We become conscious of our selfhood, or at least we become conscious of it in a new way, by becoming aware of a self who is *not us.*

Already we can see how this process is dialectical: the unreflective self is the *ideal,* or the kernel of imperfect truth that one has before experience; the Other is the *negation,* which confounds the inexperienced mind. The *actuality* arises from their conflict; it is a fuller understanding that situates the self in a world populated by other selves. Consciousness has ceased to be alone, and it has been changed forever by the encounter—the encounter, ultimately, with itself.[2]

It is characteristic of Hegel, however, that when the self learns that it is not alone, violence ensues. "[E]ach seeks the death of the other ... They must engage in this struggle, for they must raise their certainty of being *for themselves* to truth ... The individual who has not risked his life may well be recognized as a *person*, but he has not attained to the truth of this recognition as an independent self-consciousness."[3] The fundamental relationship between individuals is, for Hegel, a *fight to the death*.

If one is not already convinced, the above assertions will likely appear quite arbitrary. Why, after all, does mature selfhood require the risk of life, and possibly murder? Now, Hegel did not counsel his followers to do murder in any literal sense. The life-or-death encounter with the other has a mythic or archetypal quality to it, one that is not to be reenacted in cold blood. Still, though, Hegel held it to have been a foundational moment in the development of the human spirit, one that continues to color human society down to the present. Even in this form, it is a bold assertion, and one for which we might wish further empirical support.

To continue with Hegel's narrative, one thing the Self learns from the encounter with the Other is that some consciousnesses exist for themselves, while others are dependent: I, as a victorious self-consciousness, *exist for myself*. You, who have been vanquished, can be killed at any time. Because I have this option, I may be said to hold you in bondage.

A new conflict arises: The self-consciousness that exists for itself is the *master*; the self-consciousness that merely lives for another is the *slave*.[4] Master and slave can *only* exist in the context of one another, precisely as self-consciousness had earlier arisen only from the mutual recognition of selfhood in the presence of the Other. A new combat now pits the slave against the master. The bondsman in revolt "[rediscovers] himself by himself," writes Hegel. "[I]t is precisely in his work wherein he seemed to have only an alienated existence that he acquires a mind of his own."[5] For Hegel, it is only by passing through bondage that an *authentic* freedom can be had: Authentic freedom is, we might say, a freedom that also knows what it means *not* to be free. Note that Hegel did not say that freedom was slavery, precisely, but rather that one could not attain a particular sort of freedom without the experience, and sublation, of slavery. The two were wrapped up in one another.

Hegel believed that much of history could be explained through similar processes of conflict and resolution. Considered as a totality, and taking place across all of human history, this process constituted for Hegel a Mind or a Spirit. This Mind's workings were also the real (though seldom appreciated) subject of history. Hegel believed that when we write history, we approach excellence insofar as we capture the drama of the world spirit. The drama would end, finally, when mankind's exterior and interior lives were reconciled to one another, when the exterior compulsion of the state and the interior freedom of the individual were no longer in tension, and when the individual could see in the state the realization of his own freedom. Through many conflicts, the private wills of all individuals would be reconciled to the general will, or the world spirit, whose objective form in history was the state. The reconciliation of man and state would be of a kind with the reconciliation of self and other, and of master and slave. In time, all would be reconciled.

This was heady stuff, to be sure. But where does it leave the individual? As Leszek Kołakowski rather depressingly put it, "the practical application of [Hegel's] doctrine means that in any case where the state apparatus and the individual are in conflict, it is the former which must prevail."[6] There is good reason to believe, too, that when Hegel's difficult jargon was stripped away, this practical application was exactly his intent.

One might think, on first encountering it, that Hegel's philosophy would produce a dense, highly intellectualized history, one in which (perhaps too conveniently) philosophers and theologians would play all the most important roles. This though was not where Hegel went. Ideas always needed a concrete expression, he held. And Hegel located that concrete expression in the rise and perfection of the nation-state, which was central to his view of history.

Hegel taught that the progress of history consisted of a series of ever more authentic modes of *political* association; that is, the progress of history manifested through more and more advanced forms of statehood, while lesser and weaker states (and their attendant cultures) were gradually overcome. As each new form arose, it overcame those that had gone before it, typically with violence. Every successful civilization subsumed its predecessors, and each of them flourished—until it too was overcome by a successor. Just as in the life of the mind, in the life of politics, each moment got closer to the Absolute.

As Hegel wrote in the *Philosophy of Right*, "The state is the actuality of the ethical Idea." If we pay careful attention to the special meanings that

Hegel gave to his terms, we will find that this meant, roughly, the following: *ethics* is the bare, uninformed idea that exists before practical (political) experience. The *state* is the actual; by facing conflict, it has learned
the flaws of mere textbook ethics, and it may contradict textbook ethics
whenever it must. He continued:

> [The state] is ethical mind qua the substantial will manifest and revealed to
> itself, knowing and thinking itself, accomplishing what it knows and in so far
> as it knows it ... while self-consciousness in virtue of its sentiment towards
> the state, finds in the state, as its essence and the end-product of its activity,
> its substantive freedom.
>
> The state is absolutely rational inasmuch as it is the actuality of the sub-
> stantial will which it possesses in the particular self-consciousness once that
> consciousness has been raised to consciousness of its universality. This sub-
> stantial unity is an absolute unmoved end in itself, in which freedom comes
> into its supreme right. On the other hand this final end has supreme right
> against the individual, whose supreme duty is to be a member of the state.⁷

One might ask why other institutions besides the state are *not* actualiza-
tions of the ethical Idea, or workings of the Mind, on potentially equal
footing with the state or even surpassing it. To name just one possibility,
Augustine would have insisted that the city of God surpasses the city of
man. A man-made state could not possibly have *supreme* right against the
individual. Only God could have that.

Yet, and perhaps puzzlingly, Hegel did not believe that the non-state
parts of society could exert themselves meaningfully without the state. This
is what he meant when he wrote, again somewhat opaquely, that the state
exists "immediately in custom, mediately in individual self-consciousness,
knowledge, and activity." States *set* customs, or else they permit customs
to be, and the state's presence is immediately manifested there. We who
obey custom go on to think and act, mediately, after the way that the state
has fashioned our self-consciousness. Because we have not yet reconciled
our clashing individual wills to the state and to the substantial will, Hegel
found the realm of private life to be one of petty, inharmonious interests,
an antagonistic arena in which people realized only individual happiness
(at best), and where reason only sometimes penetrated. The idea of civil
society giving rise to spontaneous orders, or even of civil society reaching
reasonable conclusions on its own, was alien to him. Meanwhile, the state
was reason.

The state for Hegel offered the true realization of human freedom, insofar as it presented the individual with a tangible form of the world spirit. All other social institutions were at best second-order considerations. They came from freedom, but freedom came from the state. To will the existence of one's church, or school, or family, one implicitly had to will the existence of the state as well, for without it, one would lack the freedom needed to establish anything else. (A decidedly Rousseauian turn, incidentally.[8])

But where the condition of being a free citizen was for Rousseau a utopian and perhaps ephemeral ambition, for Hegel it seems to have been realized everywhere. In any actual state, an individual's supreme freedom and his supreme duty were sublated—they were united into a transcendent whole, which consisted of obedience. One's duty thereafter was to understand and appreciate the state, which was the source of one's freedom. This was because, in Hegel's view, the workings of the universal mind are in fact what we should all want, rather than our narrower aims, if ever a conflict arises—which it certainly would.

Now, one may readily doubt a premise that Hegel appears to rely on here, namely, that freedom without the state is impossible. And on the intellectual plane, one clearly can will the existence of other institutions *without* willing the existence of the state, as anarchists will immediately point out. To tell anarchists that yes, despite all appearances, they do will the state constitutes a failure to take their argument seriously.

Hegel never claimed, though, that states could do no wrong, and I do not believe that he thought this, either. In his view, many states could and *did* do wrong: After that, they were overcome by historical progress; their moment had passed. But Hegel nonetheless wrote several things that gave all states *carte blanche* in their dealings with the individual. These included, as we have seen above, the claim that the state is endowed with a "supreme right" against individuals. To this Hegel added that the state is "absolutely rational" whenever it met an all but imponderable test: As quoted above, the state is absolutely rational "inasmuch as it is the actuality of the substantial will which it possesses in the particular self-consciousness once that consciousness has been raised to consciousness of its universality." In plainer terms, we might say that whenever the state's purposes coincided with those of the world spirit in its highest level of development, the state was absolutely rational.

Hegel never quite explained, though, just how one might determine whether the criterion had been met. States exist in many different forms, as he was well aware. All had declared goals, ideologies, loyalties, and the like. Yet the *substantial* will was something different from all of these. It was a will of which states and statesmen may not even be aware. Discovering the substantial will was the whole purpose of the state's existence, but how might one might know when one had done it? A space has been opened, perhaps, for a kind of discussion, but not one in which the individual has much to say in his own defense.

At any rate, this is a safer interpretation of the passage than the simple "might makes right," which might not be fully charitable to Hegel. Indeed, Hegel writes,

> world history is not the verdict of mere might, i.e. the abstract and non-rational inevitability of a blind destiny. On the contrary, since mind is implicitly and actually reason, and reason is explicit to itself in mind as knowledge, world history is the necessary development, out of the concept of mind's freedom alone, of the moments of reason and so of the self-consciousness and freedom of mind. This development is the interpretation and actualisation of the universal mind.[9]

The writing is admittedly difficult here. While Hegel clearly blesses the verdict of history, he denies that *mere* might makes right. Better could be to say that in his view *mind makes might*, and only afterward does might make right.

Believing that the state is the actualization of the universal mind is nonetheless full of danger. Few things are so tempting to statesmen, or to philosophers, as the idea that they hold in their hands the motive powers of the world, that they are the ones who are driving history, or at least that they are the only ones who can really see what is transpiring. Like Rousseau, Hegel seems to have suffered from this delusion.

Yet even granting the existence of a world spirit, and all the rest of Hegel's system, false positives must clearly abound. Theoretically, any state could claim the mantle of the universal mind. Several states could do it at once, or even two states locked in a fight to the death, and we might only be able to guess the judgment of the universal mind by the outcome of the war. If that's not precisely "might makes right," it is nonetheless hard to see how the conversation has been substantially advanced beyond it. And any doubts we may have about the existence of the world spirit will not be assuaged by the outcome of a conflict that would in any case have had a victor. No conflict, it would seem, could possibly falsify the claim.

It gets worse: If the universal mind suffuses all of history, then the universal mind is quite a monster. *Why did the universal mind will the Holocaust?* For Hegel, the universal mind is to some degree a stand-in for God or Providence, and the theological problem of evil now looms large for the history of states just as it does for God in sacred history. When bad things happen, we may wonder how an all-powerful and all-good God can allow it. And when a state does evil, why did the universal mind permit it? Is it simply to achieve its own self-actualization? But isn't that *awful?*

In the twentieth century, the possibility of a coldly indifferent or even a malevolent universal mind has become increasingly common in philosophy, literature, and popular culture. We may find it in H.P. Lovecraft, Philip K. Dick, and some of the darker episodes of *The Twilight Zone*. In the early nineteenth century, one might still have believed that a benevolent Providence watched over us all, but this belief has become harder to affirm following the horrors of the World Wars, the Holocaust, the Great Leap Forward, and the repeated genocides of recent history. In view of these considerations, one might have a duty to oppose the state at all costs, even if the effort is doomed from the outset. If Hegel is right—and if the universal mind turns out to be cruel rather than benevolent—then anarchism is a humanism. It may be a perfectly futile gesture, but at least it is decent, and perhaps it is beautiful.

Hegel, though, informs us that in evaluating the state, we must *not* look at the history of the state's behavior, its origins, its institutional structure, or its self-justifications:

> [A]ll these questions are no concern of the Idea of the state. We are here dealing exclusively with the philosophic science of the state, and from that point of view all these things are mere appearance and therefore matters for history. So far as the authority of any existing state has anything to do with reasons, these reasons are culled from the forms of the law authoritative within it.[10]

There is a sense in which Hegel is attempting to have things both ways here: He cannot declare that state behaviors, origins, institutional structures, and justifications are "mere appearance and therefore matters for history"—while simultaneously situating the world spirit *in history*, and declaring its workings to be (1) made out of the state and (2) supremely important philosophical truths.

Foreclosed as well are ideologies based on individualism:

> In considering freedom, the starting-point must be not individuality, the single self-consciousness, but only the essence of self-consciousness … The march of God in the world, that is what the state is. The basis of the state is the power of reason actualising itself as will. In considering the Idea of the state, we must not have our eyes on particular states or on particular institutions. Instead we must consider the Idea, this actual God, by itself. On some principle or other, any state may be shown to be bad, this or that defect may be found in it; and yet, at any rate if one of the mature states of our epoch is in question, it has in it the moments essential to the existence of the state. But since it is easier to find defects than to understand the affirmative, we may readily fall into the mistake of looking at isolated aspects of the state and so forgetting its inward organic life.[11]

Only small thinkers will doubt the state. Those who are more profound will see the state for what it so often is—the living manifestation of God. Elsewhere in the *Philosophy of Right* Hegel writes,

> Immature minds delight in argumentation and fault-finding, because it is easy enough to find fault, though hard to see the good and its inner necessity. The learner always begins by finding fault, but the scholar sees the positive merit in everything … Hence men's apparent sentiment towards the state is to be distinguished from what they really will; inwardly they really will the thing, but they cling to details and take delight in the vanity of pretending to know better.[12]

This declaration is a bit astonishing, given the progress of the discipline of history: For whole millennia at a stretch, the historian's sole task was to chronicle the great deeds of the god-king or god-emperor; fault-finding was only rarely on the agenda. And I hope it will not be thought pedantic to note that during Hegel's own lifetime, authors were imprisoned, tortured, and massacred for their criticisms of the state. Perhaps these criticisms take less learning than an attempt at systematizing a whole. Perhaps. But criticism certainly took more courage and moral rectitude, and both are qualities one admires in a philosopher at least as well as erudition.

In any case, what was easy, as testified by chroniclers beginning in ancient Sumer, has almost always been to worship the state, no more and no less. A historian who wanted to keep his employment, to say nothing of his life, was well advised to be a sycophant. And at the very historical moment when it had finally—finally—become possible for intellectuals to write critically of the state before a general audience, along comes Hegel, ready to set up the old idols again. Only small minds rebel. Great ones lick the boot.

Well-developed arguments for anarchy were rare in Hegel's day, but it will not do to take even the critics of specific state actions and answer them by saying that, no, in their heart of hearts, the arguers who made these arguments still desired the state. Perhaps they did. But the proposition is quite beside the point, in considering a specific abuse.

As explained in the introduction, to my mind the state is a bundle of disparate elements. I will have more to say about this in Part II, but for the moment it should be noted that a bundle of disparate elements is not necessarily coherent, or self-conscious, or the actualization of any deeper truth at all. Perhaps it is; perhaps it isn't, and it would seem that the state's coherence must be decided on a case-by-case basis, in light of its actions. It is on the terrain of individual institutions and practices that we must always evaluate the state, and not as an image of what we desire it to be: What we inwardly will is *always* a different thing from the actual state, and it is not nitpicking to say so.

Even in Hegel's own system, states only ride the crest of world history for so long. Eventually, they cease to be relevant to history's continued march and will often find themselves playing catch-up. This by itself ought to have been a serious difficulty to his system. Hegel writes:

> The history of a single world-historical nation contains (a) the development of its principle from its latent embryonic stage until it blossoms into the self-conscious freedom of ethical life and presses in upon world history; and (b) the period of its decline and fall, since it is its decline and fall that signalises [sic] the emergence in it of a higher principle as the pure negative of its own. When this happens, mind passes over into the new principle and so marks out another nation for world-historical significance. After this period, the declining nation has lost the interest of the absolute; it may indeed absorb the higher principle positively and begin building its life on it, but the principle is only like an adopted child, not like a relative to whom its ties are immanently vital and vigorous. Perhaps it loses its autonomy, or it may still exist, or drag out its existence, as a particular state or a group of states and involve itself without rhyme or reason in manifold enterprises at home and battles abroad.[13]

Anxieties about the decline of the West, the clash of civilizations, and the like all go back to Hegel, although to Hegel himself, "the West," or liberal democracy, or some similar formation was certainly not the highest manifestation of the universal mind. To Hegel, the highest form of the state yet seen, the one in which the world spirit had actualized itself to the greatest degree, just so happened to be the kingdom of Prussia, Hegel's very own home, which was then the pre-eminent power among the German states. And if Voltaire had lived to see Hegel, he might have recognized in him the very figure of Dr. Pangloss, a caricature that seems in some ways to fit Hegel even better than it did Leibniz.

For Hegel, the history of states contained essentially four stages: the Oriental, the Greek, the Roman, and the Germanic. It is remarkable, and worth a long citation with commentary, to watch the play of early nineteenth-century prejudices in the passage where Hegel describes the four realms of state development:

> (1) The Oriental realm.
> The world-view of this first realm is substantial, without inward division, and it arises in natural communities patriarchically governed. According to this view, the mundane form of government is theocratic, the ruler is also a high priest or God himself ... In the magnificence of this regime as a whole, individual personality loses its rights and perishes ... class differences become crystallised into hereditary castes. Hence in the Oriental state nothing is fixed, and what is stable is fossilised.

The Oriental realm had no inward divisions; this claim of Hegel's may owe more to European ignorance of Asian history than to any other factor. What Hegel describes here differs little from things that Europeans had constantly told themselves about most peoples to the east of them. Aside from the patriarchial system—in large part shared, incidentally, with contemporary Europe—little of it was true.

> (2) The Greek realm.
> This realm possesses this substantial unity of finite and infinite, but only as a mysterious background, suppressed in dim recesses of the memory, in caves and traditional imagery. This background, reborn out of the mind which differentiates itself to individual mentality, emerges into the daylight of knowing and is tempered and transfigured into beauty and a free and unruffled ethical life. Hence it is in a world of this character that the

principle of personal individuality arises, though it is still not self-enclosed but kept in its ideal unity.

Now the Greeks, they had civilization. Their myths spoke of a time a lot like Oriental despotism, but individuals could make art and philosophy about them, creating "beauty," and "a free and unruffled ethical life."

(3) The Roman realm.
In this realm, differentiation is carried to its conclusion, and ethical life is sundered without end into the extremes of the private self-consciousness of persons on the one hand, and abstract universality on the other. This opposition begins in the clash between the substantial intuition of an aristocracy and the principle of free personality in democratic form ... Finally, the whole is dissolved and the result is universal misfortune and the destruction of ethical life.

The Romans pursued individualism and philosophical inquiry further, but its ends at the time were limited, Hegel elsewhere writes, because the leading schools of thought, including Epicureanism and Stoicism, had no effective politics and no way to engage with the state. In private life, Roman men became the lords of their households, which was aristocratic, and it clashed with the egalitarianism demanded by the republic. As a result, the whole civilization dissolved from its own contradictions.

(4) The Germanic realm.
Mind is here pressed back upon itself in the extreme of its absolute negativity. This is the absolute turning point; mind rises out of this situation and grasps the infinite positivity of this its inward character, i.e. it grasps the principle of the unity of the divine nature and the human, the reconciliation of objective truth and freedom as the truth and freedom appearing within self-consciousness and subjectivity, a reconciliation with the fulfillment of which the principle of the north, the principle of the Germanic peoples, has been entrusted.

This principle is first of all inward and abstract; it exists in feeling as faith, love, and hope, the reconciliation and resolution of all contradiction. It then discloses its content, raising it to become actuality and self-conscious rationality, to become a mundane realm proceeding from the heart, fidelity, and comradeship of free men.[14]

It was the task of the Germanic peoples to realize, finally, that man was in unity with God; the Germanic way consisted of faith, love, and hope, of rationality and comradeship, and of everything else that was good. In short, freedom was being a German.

Yet history did not end with Prussia, of course. The question now arises how a Hegelian can make sense of whatever comes, inconveniently, *after* the end of history: Obviously, Hegel could not have been entirely right, even if a Hegelian would by definition think that he did supply more or less the right framework. It is a question that Hegelians have struggled with ever since, and it has caused them to break into two broad camps, conservative and radical.

Let us consider the conservative interpretation first. To do so, we begin with the famous assertion in the preface to the *Philosophy of Right* that "what is rational is real, and what is real is rational." Understood as a pair of propositions in classical logic, we have a simple 1:1 identity, and a fairly uninteresting one at that.

Relaxing the constraints of classical logic, we find a sort of creative ambiguity. If, for example, we place just a slightly greater emphasis on the second half of the dictum, then the aphorism asks us to seek the rational *in* the real. Thus the leading states of one's own day can be understood as *representing* rationality; they are *real*, and thus they constitute the best possible approach to truth that is available in one's time. It follows that we should side with them whenever possible. Hegel clearly favored this view: The existing states *are* the state of the art, and even if they later prove only to have been moments in the development of Mind, we can at least console ourselves by saying that we could not have done better.

On this interpretation, Hegel's approach to history is both *conservative*, in that it does not recommend programmatic changes to the social order, and it is *stadial*, in that it looks back at previous eras as successive stages that develop toward what we have right now. Stadial theories of history were to dominate the nineteenth century in part owing to Hegel's enormous influence on the field. That influence continues today. The English Whig school of history, mainstream Soviet historiography, and the clash-of-civilizations theory offered by many neoconservatives are all stadial, and all are to some degree either influenced by Hegelianism or at least compatible with it—as well as being quite conservative ideologically, relative to their own times and places.

The case for Hegelian radicalism is somewhat more difficult to make, but it's by no means impossible. Begin again with Hegel's oracular pronouncement: what is rational is real, and what is real is rational. This time, we place the emphasis on the *first* part of the dictum: What is essential about our world—the thing that counts, the thing that is driving everything else—is the mind or the spirit as it acts in the world. All other things are less real than the mind.

And it is individuals who have minds, added the radical Hegelians. If the rational *is* what is real, and if rationality exists to overcome negation, then it is the mind's job to remake the world, and to realize a new moment of truth in the process. The mind needs to actualize itself through *critique*, and the raw material of that critique—the thing that *doesn't* drive other things—is dead, unreflective history. As it now exists, the Prussian state must be unwoven by the Mind. Perhaps even the state in general must be unwoven, as Karl Marx would go on to conclude.

My own reading of Hegel is that he was personally a staunch conservative, one who would not have favored this radical interpretation of his thought. Many passages attest to this, perhaps none so clearly as the following: "If a theory transgresses its time," he wrote, "and builds up a world as it ought to be, it has an existence merely in the unstable element of opinion, which gives room to every wandering fancy."[15] If Hegel could have seen the left Hegelians who claimed to follow him, he might have accused them of taking a ride on the wind. And yet, through Marxism, a particular breed of radical Hegelianism would come to be one of the world's leading ideologies for more than a century. It is hardly an exaggeration to say that much of history since Hegel can be understood as an ongoing conflict between conservative Hegelians, who trust the state on all or nearly all matters, and who seek to extend its power—and radical Hegelians, who want precisely the same thing, only with themselves in charge.

MARX: THE STATE IS HISTORICALLY CONTINGENT

As we have just seen, Hegelians faced a dilemma: How could one reliably infer whether a given change to one's polity was warranted? Answers were all about in Hegel's system. But these answers were contradictory, and as a result, Hegelianism was capable of satisfying virtually anyone, or perhaps virtually no one. One way of approaching the thought of Karl Marx is to consider that Marx began with this Hegelian quandary. And he proposed a compelling solution to it: Historical change is driven by mankind's ongoing encounter with the natural world, in the form of *work*, or *labor*. How people work, how work is organized, what tools exist, who owns them, and who gets the rewards of work: these are the fundamental questions of human society, and they shape or guide all the rest. Properly understood, they are also the keys to understanding the life of the mind. As Marx wrote:

> My dialectic method is not only different from the Hegelian, but is its direct opposite. To Hegel, the life process of the human brain, i.e., the process of thinking, which, under the name of "the Idea," he even transforms into an independent subject, is the *demiurgos* of the real world, and the real world is only the external, phenomenal form of "the Idea." With me, on the contrary, the ideal is nothing else than the material world reflected by the human mind, and translated into forms of thought.[16]

When historical changes in polity, culture, family life, spirituality, or even the arts arose from changes in working conditions, then these other changes could be understood as a part of the march of history. When changes happened without reference to work, they were ultimately barren, ephemeral, and not terribly important to understanding the story of humanity. As Marx wrote, "Technology discloses man's mode of dealing with Nature, the process of production by which he sustains his life, and thereby also lays bare the mode of formation of his social relations, and of the mental conceptions that flow from them."[17] Labor relations were the basis of all other social relations; in the end, their effect was decisive over all the rest, including the intellect, whose contents commonly reflected the relationships between classes in various societies.

In short, the problem inherent in Hegelian analysis was to be solved with reference to something outside of the world of the mind as it is usually understood. Mind wasn't fundamental anymore. *Technology* was, because technology shaped labor, the encounter with nature, and ultimately all of society.[18]

Classical historiography had emphasized war, politics, and rulership. As we have already seen, this approach to history placed the state at the rhetorical center of society and assigned to it a set of tasks that it could not hope to fulfill without becoming dictatorial. Political theory all too often followed suit, and Hegel's in particular proposed to use the history of the rise and fall of polities to trace the development of the human spirit.

Marx's approach to history, which he termed dialectical materialism, was by no means unreasonable as an alternative. Much indeed can be learned about cultures, and about social and mental processes, from the study of labor, and from the study of labor-related technologies in particular. Whether or not work provides the key to all else, one thing is clear: Its history, and indeed economic history in general, had been woefully neglected before Marx.

It can also readily be appreciated that the realm of work has meaningful and often decisive influence on the course of history. This observation ought not to be resisted; nor does it commit us to rest of Marx's conclusions. Indeed, perhaps it ought to have been obvious long before him. No historian today denies the importance of labor or labor technologies in shaping other social structures, and even in shaping ideas. A split, however, remains: non-Marxist historians tend to favor an eclectic approach, one in which many different types of technologies, economic trends, ideas and ideologies, and other factors all exist in a complex web of causation, with no single one of them always or primarily driving the rest. For Marxists, however, the causal arrow almost always runs from the way that labor is arranged toward everything else.

That's obviously a much stronger claim. It's also possibly not Hegelian anymore, and Marx may be fairly accused of destroying Hegelianism in order to save it. And yet—a Hegelian might have expected precisely such a reversal: Did Marx *not* perform a sublation, more or less as Hegel had described? Is it not a case of an *ideal*—a philosophy like Hegel's—coming into contact with a *negation*—the messy world of work—and the result being a new *actuality*, or a new development in the story of mankind, namely dialectical materialism?

Marx appears to have thought so. He certainly believed that the time had come—after the overly abstract and intellectual philosophy of Hegel—to *actualize* philosophy, to make it decisive in the real world. In short, the time had come for a revolution: "The weapon of criticism obviously cannot replace the criticism of weapons," he declared, and it was high time for the latter.[19] Elsewhere and more famously: "The philosophers have only interpreted the world, in various ways; the point is to change it."[20]

How did Marx know this? The state of work in nineteenth-century Europe was frankly deplorable. Early factory work was notoriously grueling, ill-paid, and unsafe. Living conditions followed quite naturally from the work, and they too were deplorable: Cramped, unsanitary, and unsafe housing became the norm in newly industrialized towns like Birmingham, Manchester, Leeds, and others. Marx's close colleague, Friedrich Engels, compiled a horrifying account titled *The Conditions of the Working Class in England* that should give pause to anyone who is tempted to regard the early phase of the Industrial Revolution as an unmitigated good. Horribly short life spans, drunkenness, ignorance, criminality, and disease were all appallingly present in the early industrial era.

Marx explained that these conditions stemmed from a single social phenomenon: *the alienation of labor.* Under the system of artisan production, each craftsman had the entire production process at his command. He typically owned all the tools he needed to do his work. He worked from the start of the project until the end, and he produced the whole of it. Work was typically done in the home, and it was often a family affair. Most importantly, the worker maintained control over the finished product, which he could sell at a price that he found just; Marx believed that workers in such a system would naturally keep the labor value that they had put into the product. If they did not, there would be no point at all to their producing it.

All of that changed under capitalism. Now workers no longer owned the tools of production; the capitalists did. Workers no longer guided a project from start to finish; instead, they performed one small and unskilled piece of the work, and they lacked the skills to do it all independently. Indeed, virtually no one had those unifying skills anymore; the machines made them obsolete.

But while this system may have been more productive in the aggregate, it reduced workers to cogs in a machine. They became both interchangeable and cheap, Marx claimed, and whenever they wore out, they were all too easily replaced. They had no marketable skills, but then again, they didn't require any. Even during their own working lives, they could take no satisfaction from their labor, whether that was material satisfaction, in the form of control over the product's sale and the profits derived from it, or psychic satisfaction, from pride in a job well done. All of this was stolen from them, or alienated, by industrialization.

Alienation was all around, Marx believed, and it was taking place both on the material plane, with the dire consequences just described, and on the intellectual one, where philosophy had become sterile and detached from reality. Alienation even affected the bourgeoisie, whom one would think were the chief beneficiaries of the whole nefarious works. But no— the lives of the bourgeoisie were dominated by petty scrambles after money. "In bourgeois society capital is independent and has individuality, while the living person is dependent and has no individuality," Marx wrote in the *Communist Manifesto.*[21] The bourgeoisie may have been relatively comfortable, but they were alienated from humanity itself.

Family, love, and communal ties were all dissolving in the search for more and more wealth, or at least in the squalid attempt to keep what one already had. Many bourgeois were declining into the proletariat, Marx held, as capital grew more and more concentrated. As a result, the bourgeoisie came increasingly to despise even the members of their own class. Some were destined for great wealth and idleness; others would be immiserated. Either way, though, the entire class was simply vanishing. The old order was dying.

The new order would arise by the simplest possible means. The proletarian class would stage a revolution, one in which the deepest, most abstract aspirations of philosophy were united with the most practical of working-class concerns. The abstract would be reconciled with the concrete; the poor would be lifted up to their rightful places in the world. As to the state, Marx declared that it too would be removed from bourgeois hands; it would become, at least for a time, a dictatorship of the proletariat, in which proletarian concerns alone were addressed.

For now, however, the state served to prop up the bourgeois order. "[T]he bourgeoisie has ... since the establishment of Modern Industry and of the world market, conquered for itself, in the modern representative State, exclusive political sway. The executive of the modern state is but a committee for managing the common affairs of the whole bourgeoisie," wrote Marx in the *Manifesto*.[22] In *The German Ideology*, he declared that the state was "nothing more than the form of organisation which the bourgeois necessarily adopt both for internal and external purposes, for the mutual guarantee of their property and interests."[23]

Government's purpose, at least for the time, was to protect the bourgeois notion of private property and to shield the bourgeoisie from proletarian agitation. This was precisely why the proletariat needed to overthrow the bourgeois state.[24] The new order, the order that Marx imagined, would do away with the bourgeois conception of private property. It would likewise do away with buying and selling, particularly of labor. The state would cease to rule the lives of men, and it would begin simply to administer *things:* The state's purpose would be to harmonize and rationalize industrial production, no more and no less. As such, no one would have reason to fear it. The state would govern all things, but no men.

Just how and why this might happen has always been less than clear to many of Marx's readers, myself included, but the idea is interesting nonetheless in the context of our investigation into the state's purposes: For Marx, the state did *not* constitute a central or transcendent principle of society. It was not especially, or even at all, concerned with virtue. Nor was it to be given a limited but ongoing charter of powers. On the contrary, the state's very legitimacy was thrown into doubt. The state was a makeshift at best, and an oppression at worst.

Moreover, the state's continued existence was contingent on remaining at a particular level of economic development. For Marx, the bourgeois state's purpose was to perpetuate the institutions that were unique to that level of development. As economic conditions progressed, we would soon see the end of the bourgeois state. Come the revolution, the proletariat would institute a harmony of personal interests with those of society; the two would no longer be at odds with one another, and the personal sphere of liberty, formerly expressed in civil society, would merge with the social sphere of liberty, formerly expressed in political action. Man would reunite his divided humanity.

Heady stuff, once again. One may still ask whether Marx's philosophy was not running far ahead of him, despite his avowed materialism. In the empirical world, Marx did do us all a great favor, however, because he made what were in effect two testable predictions: first, that the working class of Europe would grow more and more impoverished and desperate; and, second, that their desperation would prompt a communist revolution.

The second of these claims is harder to test, of course: while Marx predicted a revolution, he did not predict precisely when it would occur. Revolutions are common enough; any revolution could wear the Marxist label or have it applied in retrospect. Yet none, I think, has unambiguously passed the test. Meanwhile, the first of Marx's claims—that the working class of Europe would grow ever more desperate—is much more falsifiable, and indeed, it has been falsified. Living standards among the working class have broadly risen in all countries that have undergone industrialization, whether in Europe or elsewhere, in each case from the time that they have begun to industrialize. The working people of Europe today may have their problems, but they live a life incomparably better than the one their ancestors knew in the nineteenth century. There have been advances and reversals along the way, and one may always wish that the pace of improvement were faster. But the improvement itself is undeniable.

The mechanisms causing this improvement have been diverse. They include direct action by workers' movements, regulations promoting factory safety and better working conditions, a rising cost of labor, technological innovation, and state-provided social safety nets. We can and should debate which of these causes is pre-eminent, if any, but it is a debate that runs outside the scope of this book. Still, one point should not be lost: The misery that Marx predicted has not arrived, and we therefore have little reason to expect that his revolution will arrive either. Revolutions may claim (and many have claimed) the mantle of Marxism anyway, but often they cannot do so without significant theoretical difficulties. Signally, the two most important "Marxist" revolutions of the twentieth century—in Russia and in China—were perhaps not Marxist at all, insofar as they both took place in overwhelmingly peasant societies, which had little or no industrial proletariat to speak of. Indeed, a skeptic should ask just how these societies came to be run according to Marxist ideas at all. If we have economic determinism to thank for the broad outlines of historical progress, then these two societies are hard to explain. That they both embraced Marxism suggests that important parts of the Marxist ideology are false.

To be sure, Marxists have been aware of these difficulties. Even in 1892, when Engels was writing the preface to a new English edition of *The Conditions of the Working Class in England*, the improvement from the 1840s was impossible to ignore. But, Engels countered, the new regulations on factory work typically had the effect of concentrating capital still further; it was implied clearly enough that this would further hasten the day of capital's demise, and that a revolution was still inevitable. Over the years, various similar patches have been found to support Marxism as a going concern, and we need not examine them all here.

Whatever its practical history, Marxism's appeal to intellectuals has been hard to deny. Among that class, who would *not* want to reconcile philosophy and material reality? Who would *not* want mankind to attain—or re-attain—its authentic essence? Who would *not* want to harmonize self-interest and the public good? Who would *not* want to eliminate poverty and human suffering? And who would not want to build a new society?[25]

The Marxist meme, as Richard Dawkins might say, is incredibly prolific, and it finds its way into a wide variety of minds; above all, it appeals to the intellectuals by means of glowing promises and highly abstract language. But such glowing promises, and such abstract language, can lend themselves to a wide variety of appropriations. And even assuming perfect sincerity, acting successfully on Marxist theory in the real world is clearly a difficult proposition. Notoriously, Marx's hopeful philosophy, which proposed to reconcile individuality and society, and to produce a civilization based on scientific principles, has in practice yielded nothing of the sort. Practical Marxism has been a grotesque catalog of misery and totalitarianism.

To many, this has not been so surprising. One of Marx's earliest critics, the socialist anarchist Mikhail Bakunin, aptly observed that state ownership of the means of production, and administration of material things by so-called experts, would necessarily mean a dictatorship over people, too:

> No state, however democratic—not even the reddest republic—can ever give the people what they really want, i.e., the free self-organization and administration of their own affairs from the bottom upward, without any interference or violence from above, because every state, even the pseudo-People's State concocted by Mr. Marx, is in essence only a big machine ruling the masses from above, through a privileged minority of conceited intellectuals, who imagine that they know what the people need and want better than do the people themselves.[26]

Call it Bakunin's Throne: If you build it, and if you put anyone at all upon it, they will be a tyrant. They simply can't be otherwise; freedom for individuals cannot exist when all material things are wholly at the state's disposal, and when only a few rule at the very top of the hierarchy. Whenever the state can deny all material sustenance to a person or a group, the state's power over that group is absolute. Further, almost by sheer logic, and even without the prospect of extortion just raised, absolute power over things entails absolute power over people, because every one of our designs will at some point implicate the material world and require us to employ at least some material resources. In Marx's system, the state has the final say over all of our projects.

It was also preposterous, Bakunin noted, that forty million proletarian Germans should *all* be in the dictator class. A few of them would necessarily rule, and the rest would necessarily obey. Let the rulers come from the working class, or from any other segment of society, and it would make no difference whatsoever. Placing them on the throne would establish them as tyrants, regardless of their class background.

Bakunin's criticism of Marx was so clear, so simple, and so obviously correct that Marxists have had no choice but to ignore it. Alas, the history of Marxism in practice has been little more than the history of Bakunin's prediction coming true. It is not merely that power corrupts, and that absolute power corrupts absolutely. That would be correct, and daunting enough. But the first problem is one of logistics.

Other reasons for Marxism's failure suggest themselves as well, of course. One reason frequently advanced, particularly by Marxists, is that not all Marxists are *good* Marxists. Some are venal. Some are ignorant. Some are merely opportunists. Others may even be saboteurs. Each of these is at least superficially plausible as an explanation; such types of people exist in any sufficiently large population, and we should not expect Marxists to be any different. But, we may ask, is it really plausible that *no* cadre of Marxist revolutionaries has *ever* been sufficiently good? Could it be that they were *all* so venal as to ruin *every* revolution, all across the globe? Could it be that they all contained enough saboteurs to wreck the effort? (And just how exactly does sabotage lead to totalitarianism?) None of these explanations for the practical failure of Marxism seems terribly likely when we consider the number of times that Marxism has failed.

Here's another try, and possibly a better one: Did some unappreciated difficulty in the social sciences or in philosophy ruin the Marxists' efforts? Marxism is a tremendously subtle philosophy, after all, and the few pages I have devoted to it here are hardly sufficient for the reader to form a solid understanding of it. Much more study and reading would be necessary for any serious student, and I would dread the prospect that anyone interested in learning about Marxism would stop with this volume. As a complex philosophy, Marxism is liable to fail in unexpected ways.

One difficulty that has haunted Marxism for nearly all of its history has been the prospect of calculating just how much of various commodities are needed in various parts of the economy to achieve something like optimal efficiency. This so-called socialist calculation problem may be to blame for Marxism's failure in practice. The difficulty is not merely that efficient allocation requires repeatedly solving many thousands or millions of simultaneous equations, which would be hard enough even with modern computers. Solving the problem also involves data gathering about consumers' tastes and preferences, about what forms of tacit knowledge they possess, and about how to allocate resources in response to crises or sudden new ideas that arise during the course of a pre-set economic plan. Free-market economists have argued that markets effectively distribute collected information about these factors in the form of prices; prices in turn allow actors in a market to respond relatively rationally, and also with a high degree of flexibility, even if they don't know the exact reasons why the price of a commodity stands as it does. Collective economic planners might like to benefit from a similar system, but they are unable to do so: the sheer act of collective economic planning obliterates the price signals sent by market actors, and with them goes the prospect of conducting economic planning of the sort they desire. Much research has been done on the Soviets' attempt to set up the state administration of things, and it is fairly clear *both* that this administration was a completely intractable problem *and* that the Soviets made no serious attempts at solving it. Rather, they simply began ordering things, and people, around, not to fit an optimized plan, but rather to fit an arbitrary and highly inefficient one.[27]

Nonetheless, the Marxist meme lives on. Its appeal is hard to deny, but it seems clearly to derive from intellectual or emotional considerations, rather than from an evaluation of its performance in practice. Once again, we lack the sort of technologies that would apparently be necessary to put an otherwise appealing system into practice. Marxism promised an eventual end to material scarcity, but it did so by means of yet another skyhook, much like the Great Chain of Being was for Bossuet, or like a workable astrology was for Campanella, or like knowledge of the sacred number was for Plato. The temptation to move from "I would like this" to "This must be so" is enormous and dangerous, and I do believe that Marxism is an example of it.

Now, for Marx's own time and place—the Industrial Revolution, where every day brought a seemingly miraculous and/or alarming new machine into being—the temptation toward belief in a stable, bountiful and constraint-free world must have been powerful indeed. Given the vast new avenues of production that really were opening up, it must have been strongly tempting to imagine that all that stood in the way of a post-scarcity society was simply time, plus a bit of ill-will on the part of the bourgeoisie.

A second intellectual consideration that adds to Marxism's appeal is that it claims to achieve many philosophically desirable states of affairs in its promised social order. And yet I do believe that these purported philosophical achievements are often so ill-defined that it would be difficult to determine whether or not any given society had or had not attained them.

Consider that one may say equally well of capitalism that it achieves a synthesis between the demands of society and those of the individual: To resort to Hegelian terms, prices may be said to constitute a *sublation* of individual desire following its confrontation with the desires of others. In the form of an equilibrium price, the *ideal* of an individual desire meets with the *negation* posed by the desires and resources of others. When the consumer decides how to act, his actions constitute an *actuality*, which is nothing more than the ideal, after surviving its encounter with the negation. As a Hayekian Hegelian might say, the price of a good is a moment in the development of the Absolute, which would constitute economic equilibrium. The conflict between self-interest and social interest thus finds its resolution in the very "cash nexus" that Marx despised. In short, money would be *the symbol of exchanging value for value*—and by free exchange, the needs of the individual and the needs society would be reconciled. If the members of the society so described were capable of seeing themselves and their values realized within the market order, the sublation would be achieved in a self-conscious manner. Capitalism would be a cooperative and friendly endeavor, one entailing competition, to be sure, but neither desperation nor animosity, for each would recognize the Self in the Other.[28]

Now, my portrait of a dialectical capitalism will probably not convince very many, sketchy as it is.[29] The point in offering it is not to defend capitalism as being consistent with dialectical materialism, but rather to show that dialectical materialism can be surprisingly flabby, and that its criteria are almost trivially easy to fill. Marx's concern for the desperately poor does not admit of only one conclusion, and Marx's philosophical desires are perhaps more glittering generalities—or just games with words—than they are genuine goals in philosophy.

Yet Marx remains one of the most important figures in this book. Aside from giving inadvertent fuel to generations of state-centralizers, he is notable for defending two key ideas that have aided the skeptics as well. First, there is the idea that the state itself might be historically contingent, brought into existence not out of any innate human need, but simply because present economic conditions have caused it to be. And second, the idea that the state might one day be considered unimportant, a trivial or even nonexistent constraint on our individual and social development.

As we have already noted, beginning shortly before Marx's time, explanations for and theories about the state began to take on an increasingly economic cast, and the entire discipline of political theory was progressively invaded by economics—a development that has done much to clarify important concepts and improve our thinking about the individual's relationship to the state. One need not be a Marxist to see the appeal of these developments, of course, and to the extent that I too have been bitten by the Marxist meme, I confess that it may well be here: That the state might be a stand-in for an eventual economic solution to our otherwise intractable problems, one that we are still waiting to discover and put into practice.

Oppenheimer: Citizenship Without the State

Franz Oppenheimer (1864–1943) was a German-Jewish sociologist who held the first chair of sociology in any German university, at Goethe University Frankfurt. His most important book is titled simply *The State: Its History and Development Viewed Sociologically.* Oppenheimer is an obscure writer, particularly when compared to Marx, Hegel, or Rousseau, but he was strikingly original, and he saw clearly where many more famous writers did not.

Oppenheimer believed that states arose from conquest, and not from any sort of general agreement or social contract. On this point, he followed the lead of David Hume, who had already—and accurately—observed much the same.[30] As Oppenheimer put it, "The State, completely in its genesis, essentially and almost completely during the first stages of its existence, is a social institution, forced by a victorious group of men on a defeated group, with the sole purpose of regulating the dominion of the victorious group over the vanquished, and securing itself against revolt from within and attacks from abroad."[31]

No states had begun in any other way, Oppenheimer wrote—with the exception of a few that were or had been European colonies, such as the United States. These, though, were not a great deal better: Among the European colonies, the class to be exploited typically "import[ed] itself." That is, many from the working classes of Europe came to the New World, where they resumed being workers, and where they resumed being exploited by the political class, albeit perhaps in a more tolerable manner.[32]

Oppenheimer's claim that the state began in conquest was a strong one, but it is indeed historically true: The state is the residue of an ancient conquest. For centuries, it has replicated itself in our cultures and our mores, much like the mitochondria within our cells. But unlike mitochondria, the state has not been bent permanently to a new and useful purpose. It has not been rendered docile. The state may have legitimate purposes, but it often makes much more sense to recognize the state for the conquering agent that it was at its origin. In many ways, it remains so today.

These considerations sit badly in a political culture like that of the United States, which tends to take for granted that a social contract lies at the state's foundation, at least in our own case, and perhaps normatively in all of them. A state that was *actually* founded on social contractarian grounds could *not* be founded on conquest; the two are mutually exclusive, and this remains true whether we take Rousseau's view that the social contract was a distinct historical event, or Locke's view that contractarian reasoning arises from a consideration of present-day alternatives.

In either case, we can debate whether the United States was founded on a *valid* social contract, particularly in light of the founding documents' continued tolerance for slavery. An invalid social contract could easily serve as a cover for a real act of conquest. Or one might say that United States was (and is) social contractarian *for many*, but not for all. Excluded groups would include, as a bare minimum, slaves, Native Americans, and women, at least until individuals in these categories were fully enfranchised. Even then, one might question the degree to which their relations to the state the state can be characterized by a social contract, rather than by a papered-over act of conquest. Just as contractarianism provides a hypothetical (but possibly real) past for a just state, we might say that the conquest theory provides a hypothetical (but possibly real) past for an unjust state.

We should also consider the case of immigrants, whom Oppenheimer may have dismissed a bit too glibly. Immigrants who choose to live in a polity may express disfavor for their previous polity, at least in some respects. And perhaps they have even agreed to the contract in the new one. Such immigrants, or their descendants, have always constituted a significant fraction of all Americans, rendering the United States rather more social contractarian than less. These are useful lines of inquiry, and they deserve further attention elsewhere. Also deserving of attention is the question of the degree to which the U.S. government remains social contractarian in nature today, even as so many of its laws are now written, not by Congress, but by unelected administrators, a prospect nowhere contemplated in the Constitution.[33]

Turning our attention back to the broader world, we cannot deny that a great many states were indeed founded by conquest, exactly as Oppenheimer describes. This is a simple fact of history; think what we may about it, it will not disappear. For Oppenheimer, though, conquest was more than just a historical fact. It was also a powerful tool for analyzing present-day societies. For him, conquest did not end with surrender; a conquered population that had surrendered would typically be transformed into a *subject class*, with its conquerors taking the role of a *ruling class*.

Oppenheimer distinguished two different methods by which a person could obtain what he wanted; he termed them the "economic means" and the "political means." In the former, one worked to produce a good or a service. One then either consumed it directly or traded it with someone else. Oppenheimer made no morally important distinction between labor and commerce; both were for him a part of the economic means. The political means consisted simply of "the unrequited appropriation of the labor of others."[34] The political means was in no sense restricted to states and their functionaries; on the contrary, Oppenheimer grouped states with bandits, thieves, and swindlers, as all of them partook of the political means. Conquerors did likewise; Oppenheimer's crucial insight was to observe that the methods of conquerors did not change in any important respect following the conquest. They went right on using the political means, even as they transitioned from one-time invaders to a permanent ruling institution. "All world history," Oppenheimer wrote, "from primitive times up to our own civilization, presents a single phase, a contest namely between the economic and the political means."[35]

So what was the state *for*? At least in its early stages of development, Oppenheimer claimed that the state "had no other purpose than the economic exploitation of the vanquished by the victors."[36] This squares well with our earlier analysis of the ancient historians and political philosophers, whose role appears to have been to justify this exploitation. They generally recognized liberty only in the form of the state's own freedom from external dominion, which is not the same thing at all as individual liberty. And they vigorously defended dominion over others, particularly foreigners and slaves. In hindsight, their ideas seemingly amount to little more than a bellicose defense of the political means.

And yet the political means is ultimately parasitic on the economic means. As Oppenheimer wrote, "no state … can come into being until the economic means has created a definite number of objects for the satisfaction of needs, which objects may be taken away or appropriated by warlike robbery."[37] In short, the state did not arise to protect private property. The truth is nearly the reverse, because the state arose to prey upon it. A class could not and would not permanently devote itself to the political means if the economic means had not sufficiently advanced to make the effort worthwhile.

It is no accident that the foregoing sounds somewhat like Marxism; Oppenheimer was in part responding to Marx and had clearly accepted some aspects of Marxist class analysis. And Marx himself was responding to an earlier, classical liberal theory of class relations. Like Marx, Oppenheimer posited the existence of a productive class and an exploiter class, whose relations with one another were the guiding force of history;

like Marx, Oppenheimer's state served the interests of the exploiters. The state also took on a form that reflected the nature of the ruling class's exploitation and ensured that the exploitation would continue over various stages of history. At times Oppenheimer even approached the Marxist belief that class structure gives order to the rest of society, up to and including its ideas:

> [T]hus there starts from the fields, whose peasantry support and nourish all, and mounts up to the "king of heaven" an artificially graded order of ranks, which constricts so absolutely all the life of the state, that according to custom and law neither a bit of land nor a man can be understood unless within its fold ... [A] person not in some feudal relation to some superior must in fact be "without the law," be without claim for protection or justice.[38]

The transition from feudalism to industrialism had concentrated the political means in fewer hands. In some ways, it had therefore caused the state's power to grow. With the spread of money and the demise of in-kind payments, Oppenheimer wrote that "the central government becomes almost omnipotent, while the local powers are reduced to complete impotence."[39] The political means had become concentrated in one institution, the state, which had suppressed all rivals.

But in the concentration of the political means, a contradiction is emerging in Oppenheimer's account of history: Where formerly almost anyone could use the political means, it now becomes difficult for any but state actors to do so. When so restricted, the political means looks increasingly illegitimate, because its distribution is unequal. This again resembles Marxism, which predicted that the oppressor class—for Marx, the capitalists—would grow fewer and fewer in number, even as their power grew more consolidated, and their security in power declined.

Looking to the future, Oppenheimer believed that one day the political means would be abolished, much as capitalism had abolished feudalism, and much as the proletarians would in the Marxist future abolish private property. The state as we know it, of course, would cease to exist, and it would be replaced by a social order that Oppenheimer termed "freemen's citizenship." Opaquely, Oppenheimer wrote that "the bureaucracy of the future"—not, mind you, a *state* bureaucracy—"will truly have attained that ideal of the impartial guardian of the common interests, which nowadays it laboriously attempts to reach."[40] Private property would clearly still exist, for property is a necessary consequence of the moral injunction—

which Oppenheimer subscribed to—that laborers must always be allowed to keep the product of their labor.

Oppenheimer broke with Marxism most clearly in this embrace of private property, and in his identification of an individual's social class not by his relationship to the means of (economic) production, but by whether or not he employed the political means. In both of these, he was drawing on a tradition of class analysis that long predated Marx and that continues among classical liberals even to the present day. This form of class analysis does not view control over the means of production as the origin of class exploitation; rather, class exploitation comes from the control of the state, because, in Oppenheimer's terminology, in the modern era the state is the locus of the political means. The political class is the class that can get the state to do its bidding. The economic class, meanwhile, encompasses all those people, *whether rich or poor*, who make their living without relying on unrequited transfers.[41] As a result, there is nothing intrinsically "left" or "right" about class analysis, and there is certainly nothing that about it that commits us to all of Marx's other methods or conclusions.

Class analysis of any type remains risky, however. It can lend itself to greedy reductionism and hasty generalizations. To hazard an example from recent politics, Republican presidential candidate Mitt Romney drew considerable ire when he declared:

There are 47 percent of the people who will vote for [Barack Obama] no matter what. [They] are dependent upon government ... believe that they are victims ... believe the government has a responsibility to care for them ... believe that they are entitled to health care, to food, to housing, to you-name-it ... These are people who pay no income tax.[42]

Superficially, Romney's comment bears a strong resemblance to Oppenheimer's account of the state: A political class makes its living by forcing others to pay taxes. But things are more complex than Romney suggested. In our society, as opposed to Oppenheimer's model, nearly all adults go through at least one phase of life, if not several, in which they pay no income taxes; the "47 percent" are typically *us*, whether now, or formerly, or at some time in the future. Nearly all of us receive government benefits without paying income taxes ... sometimes. Party affiliation has little to do with it, and class affiliation perhaps little as well. The political means has been dispersed throughout the citizen body.

This is not to say, however, that we are all equally of the political class, or that the term has no meaning, or that Oppenheimer's model is useless. Some in our society are vastly more a part of the political class than others, and among those who have received government benefits, we can draw many meaningful distinctions. Some uses of the political means may even be legitimate.

But on whatever theory we crafted, clearly not all uses of the political means would survive scrutiny: Whereas emergency medical treatment for the indigent seems a relatively defensible use of unrequited transfer, the public funding of profitable sports stadiums does not. This funding helps only those who are already exceptionally well-off, and it does not assist them with anything that is a necessity to their lives. Short of freelance thievery, a purer example of Oppenheimer's political means would be difficult to find. We may debate which types of help are legitimate, if any, but we should not lose sight of the fact that our society still suffers from an excess of the political means, if judged by almost any reasonable standard. In many ways, our state is still the residue of a conquest.

Oppenheimer's predictions of a coming anarchy were based on a simple extrapolation of what he believed was a historical trend. In the earliest times, "the right to the economic means, the right to equality and to peace, was restricted to the tiny circle of the horde bound together by ties of blood ... while without the limits of this isle of peace raged the typhoon of the political means."[43] Trust was weak, and the defensive use of the political means was often the only thing keeping a given group alive—for they inevitably had to contend with the political means being employed by a neighboring group.

Gradually, though, the islands of peace have expanded, Oppenheimer claimed. They had grown from small circles of barter among family and clan members, to barter among neighboring clans, to money and credit, and to international trade. Although one could always try to make a living by the political means, and although many still do so today, the aim of civilization has constantly been to make this form of livelihood less and less attractive, less and less viable. The political means has now been driven into a small segment of society, one known as the state, and from there, it will eventually be eliminated.

From a horde of conquering bandits, into to a horde of stationary bandits, the state has gradually evolved into the protector of the economic means itself. Again, we notice a resemblance to the work of Mancur Olson, noted earlier, on the progress from banditry to statehood.[44] This process could only continue, Oppenheimer thought:

> Merchants' law becomes city law; the industrial city, the developed economic means, undermines the feudal state, the developed political means; and finally the civic population, in open fight, annihilates the political remnants of the feudal state, and re-conquers for the entire population of the state freedom and right to equality.[45]

Yet anyone wishing to follow in Oppenheimer's footsteps must insist very specifically on the importance of the political means retreating from individual lives, and not merely the state retreating. Sometimes, when the state retreats, the political means actually advances, in that its unrestricted use becomes more common or more fearsome. The political means can always bear some other name, or some other organizational structure, and without the state, these formations may become more likely.

At various times in history, the so-called *intermediate institutions* have been quite domineering and have severely restricted individual liberty. Clans, tribes, and religions have all played this role, including suppressing the crucial right to exit from the jurisdiction of the group. They have done so even within what are ostensibly fully realized nation-states. The Middle East presents many examples of this phenomenon, even in areas where no state breakdown has occurred. At other times, though, intermediate institutions can provide important checks against the power of the state, and against one another. If the competition can be conducted peacefully, the sheer presence of competing intermediate groups may also conduce to the banishment of the political means—because one obvious way that groups may compete for membership is by promising a refuge from predation. As a result, our analytical work is not done simply by inquiring whether a group is a state group or a non-state group; either may potentially do good or ill, in Oppenheimer's terms.[46]

For all forms of liberalism, that is, for all political theories that take individual liberty to be the aim of politics, the real problem is not the state. The problem is the arbitrary power to command the lives and labors of others. For liberals, the state at its best is a solution that has been offered to the problem of the political means. At its best, the state aims to build a cage—made of power itself—whose sole potentially legitimate purpose is to contain power.

The liberal state may therefore be thought of as a *kludge*—an inelegant, workaround solution to the otherwise intractable problem of banishing the political means. It seems clear to me that we liberals don't know how to eliminate the political means altogether. (We as a society may also lack the will, of course, but that's another question.) The kludge that is the liberal state makes use of the very thing that it aims to check.[47] As a direct result, the liberal state remains dangerous. In the eyes of many, the liberal state is a kludge that has failed and that should be abandoned. Oppenheimer believed that for the moment we could do no better, but he also held that one day humanity would no longer need the state in any form at all. The circle of trust would grow and grow—from its origins in the family, up through commercial society, until one day there would be no reason, and perhaps no inclination, for anyone to use the political means against anyone else.

I am uncertain whether to share Oppenheimer's optimism or to condemn it. A strong empirical case can be made that the world is in no sense moving toward statelessness anymore, if it ever was. But in itself this does not mean that statelessness is an impossible or an undesirable goal. As Oppenheimer and many other classical liberals understood it, there are serious moral objections to the use of the political means, and the attempts to defend its use philosophically are much weaker than most in the political mainstream would care to admit.[48] The fact that states have nonetheless historically predominated over stateless societies suggests that states exist for a reason, but it does not necessarily suggest that they exist for an ethically defensible reason.

Today's anarcho-capitalists propose to replace the state with a variety of other institutions that would perform nearly equivalent functions, albeit without recourse to the political means. At least nominally, these institutions would be private. As to size and scope, they might be termed intermediate institutions, although they would have no state above them. They would offer protection and/or arbitration services, and they would compete with one another for clients, each of whom would pay voluntarily for as long as they wished to continue subscribing to the service. No involuntary taxation or other exactions would occur.

Could anarcho-capitalism work? I suspect that private protection agencies, like the state, would also be kludges. They too would be imperfect, and insofar as they resorted to unrequited transfer, they would be using the political means. I further suspect that the great libertarian theorist Robert

Nozick was completely right when he predicted that anarcho-capitalist protection agencies would inevitably coalesce—or maybe devolve—into power structures that are indistinguishable from modern states.[49] Whether this development should be applauded or condemned is a question independent of the matter of its likelihood.

Nozick's claim that anarcho-capitalist protection agencies would coalesce into states has touched off an ongoing and complex debate, one that we cannot explore in detail here. I bring it up only to observe that even among modern libertarians, the central problem of social theory has often been to justify the existence of the state, or else to explain why, despite all appearances, no state should be. In either case, the state remains central to their thinking, as it has been to the thinking of the discipline of political theory more generally. But why should *this* be? And what would social theory look like if the state were *not* treated as the central subject of inquiry, but rather as one tool among many for solving a set of difficult problems? These are the questions that we will consider in Part II.

NOTES

1. Hegel, *Phenomenology of Spirit*, para 20, p. 11.
2. Ibid., paras 178–186, pp. 11–113.
3. Ibid., para 187, pp. 113–114.
4. Ibid., paras 188–89, pp. 114–115.
5. Ibid., paras 195–196, pp. 117–119.
6. Kołakowski, *Main Currents of Marxism*, p. 63.
7. Hegel, *Philosophy of Right*, https://www.marxists.org/reference/archive/hegel/works/pr/prstate.htm.
8. For some further connections between Rousseau and Hegel, see Jeffrey Church, "The Freedom of Desire: Hegel's Response to Rousseau on the Problem of Civil Society," *American Journal of Political Science* 54 (1), January 2010, pp. 125–139.
9. Hegel, *Philosophy of Right*. https://www.marxists.org/reference/archive/hegel/works/pr/prstate.htm#PR342.
10. Hegel, *Philosophy of Right*, https://www.marxists.org/reference/archive/hegel/works/pr/prstate.htm.
11. Ibid.
12. Ibid.
13. Ibid.
14. Ibid.

15. Hegel, *The Philosophy of Right*, preface. https://www.marxists. org/reference/archive/hegel/works/pr/preface.htm.

16. Marx, Afterword to the Second German Edition of *Capital*, available at https://www.marxists.org/archive/marx/works/1867-c1/p3.htm.

17. Marx, *Capital*, vol I, ch 15, n 4, available at https://www.marxists.org/archive/marx/works/1867-c1/ch15.htm.

18. The place of the mind in Marxist historiography is complex, and only the most vulgar Marxists would consider mind a mere epiphenomenon. But this raises questions about whether they have truly escaped the chicken-and-egg problem in Hegelian accounts of historical causality: Is the real the rational, or is the rational the real?

19. Marx, "A Contribution to the Critique of Hegel's Philosophy of Right," at https://www.marxists.org/archive/marx/works/1843/critique-hpr/intro.htm.

20. Marx, "Theses on Feuerbach," in *Marx/Engels Selected Works*, Volume One, pp. 13–15. Available at https://www.marxists.org/archive/marx/works/1845/theses/theses.htm.

21. Marx, "Manifesto of the Communist Party," in *Marx/Engels Selected Works*, available at https://www.marxists.org/archive/marx/works/1848/communist-manifesto/ch02.htm.

22. Ibid., https://www.marxists.org/archive/marx/works/1848/communist-manifesto/ch01.htm#007.

23. Ibid., "The German Ideology," in *Marx/Engels Selected Works*, available at https://www.marxists.org/archive/marx/works/1845/german-ideology/ch01c.htm.

24. Ibid., https://www.marxists.org/archive/marx/works/1845/german-ideology/ch01d.htm.

25. Yes, there are philosophical anti-essentialists, who deny that mankind has any such thing as an essence at all. And the skeptics on this point have become much more common in the years since Marx. But if for the sake of argument an essence could be postulated, well, wouldn't it be terrible if we'd somehow lost it?

26. Bakunin, "Some Preconditions for a Social Revolution," in *Bakunin on Anarchy*, p. 338.

27. Boettke, *Socialism and the Market*.

28. One should recall Kant's otherwise enigmatic claims about the "unsocial sociability" of mankind. I think also of Jason Brennan's *Why Not Capitalism?*, an elegant response to G.A. Cohen's *Why*

Not Socialism? Cohen proposes that capitalism springs essentially from antagonistic feelings, and that these feelings are in many contexts alien to all decent people; why then should they not be rejected at all times? Brennan argues that capitalism need not be incompatible with friendship and fellow-feeling, and that at any rate socialism is something wholly different from the social instantiation of niceness.

29. The idea came to me while reading Kołakowski's *Main Currents of Marxism*, particularly book I, chapter 13. At reading, on page 250, that "In the class-consciousness of the proletariat, historical necessity coincides with freedom of action"—I was struck that there is *another* plausible coincidence of historical necessity and freedom of action, namely the market. The market arises *of* various human necessities, and action in the market consists of the attempt to provide for them. But individuals in the market nonetheless possess a type of freedom. I found reinforcement of the idea in the conclusion to the chapter, in which Kołakowski wrote that "by cancelling the legal right of ownership [socialist polities] have destroyed the machinery which gave society control over the product of its own labor" (p. 274).

30. Hume, "Of the Original Contract," in *Essays*, p. 465.

31. Oppenheimer, *The State*, p. 13.

32. Ibid., pp. 17–18. Curiously, he did not mention the importation of slaves, which would have strengthened his case. Slaves certainly did not import themselves.

33. For the case that the United States was founded on a sufficiently contractarian basis, see Amar, *America's Constitution: A Biography*.

34. Oppenheimer, p. 25.

35. Ibid., p. 27.

36. Ibid., p. 15.

37. Ibid., p. 27.

38. Ibid., p. 223.

39. Ibid., p. 243.

40. Ibid., p. 275.

41. Palmer, "Classical Liberalism, Marxism, and the Conflict of Classes."

42. David Corn, "SECRET VIDEO: Romney Tells Millionaire Donors What He REALLY Thinks of Obama Voters."

43. Oppenheimer, p. 279.

44. Mancur Olson, *Power and Prosperity*, notably pp. 6–9.
45. Oppenheimer, p. 280.
46. Two exceptional books have recently been published that examine this danger. They are Weiner's *The Rule of the Clan* and Levy's *Rationalism, Pluralism, and Freedom*. The problem of intermediate organizations—those that are between the state and the individual in scope—is a serious one for any form of liberalism, and it is particularly serious for modern libertarianism.
47. In *Federalist* no. 51, James Madison was particularly aware of this danger: "In a society, under the forms of which the stronger faction can readily unite and oppress the weaker, anarchy may as truly be said to reign, as in a state of nature, where the weaker individual is not secured against the violence of the stronger."
48. For a clear, well-argued treatment of this subject, see Huemer, *The Problem of Political Authority*.
49. Nozick, *Anarchy, State, and Utopia*, pp. 88–119.

PART II

Toward a New Theory of the State

CHAPTER 8

The Structures of Political Theory

Much like the central nervous system, the "central" state owes its adjective to its authority, not to its literal position. Calling the state central is a rhetorical component of many of the theories that we have just surveyed. Central positions symbolically assert authority and command. Even in the liberal tradition, many writers have placed the state rhetorically at the center of society. So we read the following in Thomas Paine:

> That which is called government, or rather that which we ought to conceive government to be, is no more than some common center in which all the parts of society unite. This cannot be accomplished by any method so conducive to the various interests of the community, as by the representative system. It concentrates the knowledge necessary to the interest of the parts, and of the whole. It places government in a state of constant maturity … and is superior, as government always ought to be, to all the accidents of individual man, and is therefore superior to what is called monarchy.[1]

The above is hardly a representative passage of Paine, but when he sought to defend representative government—for a good liberal reason, as a check on arbitrary power—he *still* reached for the rhetoric of centrality. It is astonishing how often and how casually writers will make this move, even when they would not otherwise assign the state the role that a central position implies.

I began writing this book in part as an exploration of what seemed like a strange rhetorical tic, but the project quickly grew into much more

© The Author(s) 2017
J. Kuznicki, *Technology and the End of Authority*,
DOI 10.1007/978-3-319-48692-5_8

than that. I could not help but notice, as others have before me, that the proponents of expansive state power have been many, while those who have offered relatively modest political philosophies, like Oppenheimer or Locke, have been few.[2] Worse, it would seem that the judgment of history itself is skewed: The best of the authoritarians—and some are indeed profound—have *always* made their way into the western canon. But many far less than lucid thinkers have ended up there as well. Crimes against clear writing, or even against logic itself, would appear far less damning than crimes against the sanctity of the state.[3]

Now, it could be that the majority has observed the actual state of things, and meanwhile the dissenters have failed to see the obvious. But I do not think that this is so. The fact that political philosophers have so often reached for what we now recognize as fantasies—for now-discredited theories like numerology or astrology—should suggest that the bias lies in human beings, not in reality. People across many times and cultures appear overly eager to arrive at roughly the same set of conclusions, by hook or by crook. These conclusions include the idea that society should be effectively ordered by a central orderer or ordering principle, that achieving such a state of affairs will bring either perfect or near-perfect stability, and that such stability is preferable to all or nearly all other potential social patterns.

Note that while all three of these conclusions are staples of the canonical texts of political theory, they all lie far outside the present political mainstream. For this we should all be thankful. One would be hard pressed to infer the ascendance of classical or even modern liberalism if all one had to go on were the canonical texts of western political theory, in which a central orderer almost necessarily dominates everything around him. From these texts, one might have predicted the rise of communism or fascism, but one could not have predicted their fall, or that they are now generally regarded as having been horrific wrong turns. By the standards of the canon, most everyone in the United States, and even in Europe, is just this side of being an anarchist. Not just the libertarians, but nearly all of us.

Even in areas where nonintervention today enjoys a broad consensus, past actors almost always reached for a state solution. Following the Reformation, for example, the ideas of religious tolerance, of church-state separation, and of acceptable religious pluralism were all incredibly slow to develop. Rulers came and went; with them came different official religions; with the official religions came alternating waves of persecution. The English crown first persecuted the Protestants, then the Catholics, then the Protestants again, and then it persecuted the Catholics once more. For well over a century, the solution proposed by each side was to try to convince the next monarch to persecute the *other* group. And never to leave matters alone. State control was simply that hard to let go of.

In this and in numerous other areas, it would seem that by our collective lights, the great majority of political theorists and practitioners have gotten things wrong constantly, all across western history. Admittedly, perhaps *we* are the ones who are wrong, but let us continue to assume we are not. We might now attempt to analyze what seem to us the philosophical errors of past. These are not random errors, for as we have seen, they tend to point in a common direction—toward a government whose powers are greater than what we now know, and, in theory, toward a government stronger than any yet seen on earth.

The errors are systematic; indeed, they are best understood as *one* error. So how does this error arise? Why would so many philosophers—in so many different times and places—all seem to converge here?

Some of the forces that have bent philosophy in this direction are readily understandable. They are what historians term *structural* forces. This doesn't mean that they are logically valid reasons—rather, it means that they arise from the economic and social conditions that prevail in a given society.

For example, we can immediately understand why Plato was appointed court philosopher to Dionysius of Syracuse, in preference to someone like Diogenes the Cynic. Plato told Dionysius almost precisely what he wanted to hear; Diogenes did not. Structural forces were also clearly at work in the case of Bossuet, whose absolutist political theory obviously furthered his career as a court intellectual to an absolutist monarch. Contemporary dissenters would never have had the same opportunities, not even if they had had intellectual gifts to match. Or consider Campanella, whose life path was much grimmer than Bossuet's. Campanella nonetheless wrote many of his works, including *The City of the Sun*, in the hopes of getting himself out of prison and into a comfortable, high-paying sinecure. And it is hard even to imagine Machiavelli without the formative experience that patronage gave to his life.

Whether one had the ear of a king, or whether one merely hoped for it, the incentives all ran in the same direction: Monarchs have almost always wanted to be told that they ought to be all-powerful. Those who told them what they wanted to hear could rise to the top. In most eras, one rarely got the chance to write history or political theory unless one was already sympathetic to the frequent use of arbitrary power. To gain or keep a position as a court author, it pays to flatter. And even long after a given author's death, history tended to be kinder if he had supported the power of the state. Those tasked with praising and preserving books—or with condemning and destroying them—generally faced the same structural imperatives as the authors themselves.

Particularly before the era of print, making a living as an intellectual apart from the state, and apart the church, was all but impossible. Relying on state preferment was almost the only way to live the life of the mind, at least while writing on secular topics (It is interesting that two of the earliest significant critics of arbitrary power, Augustine and Gratian, were both churchmen in good standing, and not heretics, like Campanella). Independent political theorists have certainly existed, particularly in the modern era, but overall they have been fewer in number, and they have typically gone to school at the feet of the statists, either literally or figuratively: Marx had to shift for himself as an impoverished journalist, but he had Hegel for a muse, and Hegel had Prussia to support him. The results speak for themselves.

Besides the problem of state preferment, we should also consider the subjective experience of *writing* and *reading about* a meticulously planned community. One plausible reason why authoritarian utopias have been so popular may be that for people of a certain cast of mind it feels good to write them. And of course, for a corresponding cast of mind, it feels good to read about them. If authors and readers derive good feelings from tales of authoritarian utopia, then we will encounter more tales of authoritarian utopia than would empirically be justified. But liking the sound of a thing does not render it a workable model for society. Thus despite Plato's frequent condemnations of poets, whom he banished from his utopia, Plato himself may stand accused of being a poet in the worst sense of the word, that is, of being a man who favors sound over truth.

It's time for a confession of sorts. Authors—myself included—constantly face the temptation to declare that we have figured it all out, and that the last thought has finally been thunk. By us! Those who succumb, even momentarily, will experience a rush of euphoria like few others on earth. It truly feels good to write in this mode. And it doesn't take any particular genius to do it, either. I have read many accounts of utopias that were so sloppy and ill-considered that I have pitied their authors, who could not possibly have written them from any genuine insight, and who must have proceeded only on feeling.[4] This emotional temptation tends to bias us, I believe, toward visions of society that are neater and tidier than they ought to be, toward accounts that leave no loose ends hanging. When we read, we ought perhaps to mistrust accounts of this type more than we usually do.

Admittedly, I advance some ambitious theories of my own in this book. But I must stress that I consider them provisional, and I would urge you to consider whether I might be mistaken. Please, I would ask, consider that I might be mistaken *particularly* if my book makes you feel really, really good inside. Feelings of exactly this type, shared between author and reader, seem likely to have led the entire discipline of political philosophy systematically astray for much of its history. Do not trust them.

The subjective experience of past readers seems likely to have pulled toward utopia as well. The past was often a miserable place, and not only for perennial prisoners like Campanella. Even the powerful led lives full of uncertainty and pain. War, plague, or famine could strike at any time. Demons haunted the imagination, and even common maladies had neither cures nor even valid explanations. Entertainments were few and expensive. The promise of a perfect and permanent social order, one from which all evil had been banished, must have glittered all the more brightly. Reading a persuasive depiction of a perfect world must have been a blissful escape, even if one could not personally hope to enjoy absolute authority, and even if the image one beheld was not of one's own making. Escapism, or, more charitably, the earnest wish for something better, may explain why utopian theories, even bad ones, have at times acquired fervent followings.[5]

Finally, for writers and readers alike, texts become more agreeable or favored when they have particular plot structures, and one reason why the utopian tendency has been so popular may be that it lends itself especially well to agreeable plot structures. Yet there is no reason why we should expect agreeable plot structures to track workable social systems, and thus the demands of emplotment may bias political theorizing: We should be wary of political theories that tell especially satisfying stories, including but not limited to those with a distinct beginning, middle, and end; those that promise events and personages infused with symbolic meaning; those that invoke heroes and villains much more unambiguous than those found in ordinary life; and those that tell of a conflict resolved in a markedly permanent and satisfying manner.

In short, I infer that the subjective experiences of writing and reading about political and social theory are real factors to be weighed in the history of these disciplines. Agreeable subjective experiences have likely skewed political philosophy away from political modesty. Agreeable subjective experiences may inspire many other intellectual projects, and the scope of applicability for this idea may turn out to be quite large.

Now, explanations like these may account for the prevalence of authoritarianism in the western political canon, but they do *not* establish that authoritarianism is false. I do not argue *ad hominem*. I mean only to point out that whether these arguments are right or wrong, we can account for why they lean the way they do, and our account does not give us any reason to find that they necessarily track the truth.

One may pin similar structural reasons on me, of course, and on the arguments that I advance. I work for the Cato Institute, a libertarian think tank. The Cato Institute advocates a much reduced role for government in society. We stand for classical liberalism and free markets, and I could not expect continued employment if I ever declared that my sympathies lay with Marx or Rousseau. But all by itself this hardly means that I must be wrong. I could be bought, and paid for, and still be annoyingly correct. To be fair, the same may be true of any of the meticulous utopian planners in the western canon. There is perhaps no significant political perspective in existence, whether true or false, that has not at some point been bought and paid for.

But I think it's probable that the strongest arguments for authoritarianism have already been articulated. Brilliant, well-funded intellectuals have toiled over them for centuries, with the aid of centuries of authoritarian practice. While it is true that no practical authoritarianism has ever approached the societies described in some of the more strictly authoritarian political theory, one would still expect real-world experience to count for *something*. Authoritarianism is to all appearances a direction that has been thoroughly explored.

On the other extreme, I do not believe that we have yet heard the best cases *against* the authority of government. Nor have we had the kind of experience that would test such theories as fairly as we might like, in all areas of human action. It remains a relatively new idea that governments might let go of the presumption of control, and that if they did so, society might be better off. Political theory has paid too little attention to the possible opportunities for doing without the state.

Although structural and psychological forces can explain the authoritarian bent of western political philosophy, I do not believe, as Marxists seem to believe, that outlining these structural explanations is the only valid way of considering important trends in the history of political thought.

If I did, I might be either unconcerned, or perhaps fatalistic, about the authoritarian bent of political theory. But there would be little I could actually do.

I do believe that structures matter, but I believe that what we *think* matters as well. The content of our minds is worthy of attention not simply because it reflects the shape of our material lives, but because our ideas directly influence the types of societies that we are willing to tolerate, experiment with, and defend. And canonical political theory, at least, pushes us powerfully in a particular direction: If we in the present day are considering an expansive project to be undertaken by the state, we will inevitably find that it is still relatively modest when compared to the broad sweep of the canon. Political theory stands ready to urge us onward in *any grand project at all*, and to suggest that even our most enormous state-run endeavors are not to be feared. Voices urging caution have been fewer and less persuasive.

It matters, meanwhile, that many of these political theories have been scientifically groundless, even when they have claimed the name and the trappings of science. There is good reason to think that the entire discipline of political theory has never had the kind of social scientific foundation that would make political engineering possible. It has not had the knowledge that it would need to achieve the vast projects it has proposed, and perhaps it never will. Meanwhile, political theorists have happily filled these gaps with appealing but dangerous nonsense.

Can the gaps ever be filled with something better? That is, can they be filled by something more convincing than numerology, or astrology, or "scientific" socialism? This is a troubling question. Beginning in the seventeenth century, the growing success of physics at explaining the observed physical world gave rise to a burst of optimism in many other fields, including economics and political theory: If laws had been found to explain motion and gravitation, heat and work, gasses and liquids, and so much else—well then, what about laws that to explain people and societies?[6] We have set aside Hobbes' peculiar physicalist account of human motivation, but what matters here is that he did have one, and he expected that it would undergird the rest of his system. Much social thought since his time has proceeded along similar lines; if the physicalism has been less crude, still, theorists have not blushed to speak of inexorable laws and processes of history.

In a sense it is hard to condemn such a project. In light of the promised payoffs, one might certainly want it to succeed, as Marx clearly did. And yet even data gathering in the social sciences is vastly more difficult than data gathering in the physical sciences. Human motivations and mental states remain to a high degree inscrutable. Even when considering individuals' external actions, social science data is necessarily spotty. It turns spottier still when we are forced to ask individuals directly about their motivations—they may lie; they may fail to fully articulate a complex motivation; they may even be unsure of just why they took a particular action. Yet without reliable data in this area, gaps will inevitably remain in any social theory. And gaps certainly remain today, despite the long search for the types of fundamental, empirically supported laws that would allow us to do useful social engineering.

Consider the discipline of history. As it supplies much of the raw material for political theory, history must shoulder a good deal of the blame for political theory's own poverty. For nearly all of its existence, history has concerned itself chiefly with kings, dynasties, wars, palace intrigues, and the various other deeds of the state. Much less has it asked about outside factors that influence the state, and almost never has it asked about stateless spaces—which would, if one were being scientific, presumably serve as the necessary control in the experiment. The state has commanded nearly all of historians' attention. Factors explaining its action have generally been taken to be endogenous, or nearly so, to the state itself; and this may be the result of nothing more than a selection bias.

Only occasionally have the histories of other parts of life been given the sustained treatment that they have deserved; only occasionally have these other parts been considered as possible driving forces in their own right. The state's closest rival in this respect has been the church, but perhaps this was again a function of the church's ability to direct patronage, and to employ the political means. Karl Popper wrote memorably of history's pro-political bias in *The Open Society and Its Enemies*:

> [T]he realm of facts is infinitely rich, and … there must be selection. According to our interests, we could, for instance, write about the history of art; or of language; or of feeding habits. … [But] what people have in mind when they speak of the history of mankind is, rather, the history of the Egyptian, Babylonian, Persian, Macedonian, and Roman empires, and so on, down to our own day. In other words: They speak about the *history of mankind*, but what they mean, and what they have learned about in school, is the *history of political power*.

... It is hardly better than to treat the history of embezzlement or of robbery or of poisoning as the history of mankind. For *the history of power politics is nothing but the history of international crime and mass murder.*[7]

This focus on the state as the truest or most important part of history arose for structural reasons, which we have discussed above. But it has systematically warped our view of what mankind has been. And it has almost certainly stunted our view what mankind could be.

Describing the state as central to the human condition ought to appear as a strange way to think. Certainly it is not how we think about the state when we are not actively thinking about politics. Few other than civil servants will wake up every morning and ask themselves how they might best serve the state. Few will choose their religious faith, career path, or mate based on what these choices might mean for the state. And virtually no one will lie on their deathbed thinking "All those years I really should have paid more in taxes."

Happily, the biases of historiography appear to be diminishing; history has gradually been approaching the way that non-theorists have probably always thought about the state, which is to assign the state a smaller role and to make themselves less subservient to it. Even in his own time, Popper's indictment of history was slightly antique, and now it may be less true than ever.

The changes began in the eighteenth century, when the bourgeoisie was finally wealthy enough to support the work of independent historians and social thinkers—individuals like David Hume, Adam Smith, and Edward Gibbon. The bourgeoisie achieved this feat in part through the university system, of which they were increasingly the patrons, and in part through the commercial book trade, which the bourgeoisie had created. Bourgeois history, if we may call it that, has concerned itself to a much greater extent with economics, popular culture, science, family and children, and the lives of women, who have typically been excluded from state power—and who thus rarely made history as it was formerly told. This bourgeois history has also frequently been skeptical of state power. It could afford to be, as its patrons were financially independent of the state, and their interests did not always coincide with the aggrandizement of state power (And as we see in Locke's case, sometimes patrons could be revolutionaries).

A good deal of the impetus for this dramatic change in historiography also came from Marxism, which declared that economic history held the keys to all the rest, and that the state was a consequence rather than a prime mover in history. Now, of course, nothing about studying economic history commits one to Marxism, and even non-Marxists can also derive much insight from Marxist historiography. Indeed they should: Marxist historiography is an important part of what has transformed the study of history from an account of the state and its crimes into something more and more like a full narrative of the human story. For this at least, Marxist historians deserve thanks.

In our own time, academic history has grown unbelievably diverse, even to the point of ridicule by conservatives, who may not appreciate that it is a sign of a culture's strength, and not its weakness, that it can devote resources to the study of early American midwives, or transgender people in the nineteenth century, or the emergence of new forms of urban slang. In part, these new types of history are just good, if sophisticated, fun. It can be fascinating and deeply rewarding to read about other human beings who have had lives radically different from (or similar to) one's own. But these new histories also tell stories that have long gone untold, and perhaps not for any good reason. Perhaps these histories may rouse political theory from its slumbers, and show it that there are more things in heaven and earth than are dreamt of in its philosophy.

Let us consider some examples. The historian and philosopher Michel Foucault began his life as a Marxist, but he ended it as something much closer to an individualist anarchist. He even recommended studying the works of Ludwig von Mises and F.A. Hayek, and he explored, in works like *Discipline and Punish* and *Madness and Civilization*, the links between the state, the sciences, and control over individuals' bodies and liberties. Journalist and historian Brian Doherty has written the story of a wholly new kind of intentional—and nearly state-free—community in his book *This Is Burning Man*. Anthropologist James C. Scott is the author of *Seeing Like a State*, *The Art of Not Being Governed* and *Two Cheers for Anarchism*. In them, he explores both the limits of state power and the strategies that individuals across many cultures have deployed to evade the state—and to create decent institutions anyway. George Chauncey's *Gay New York* revealed a subculture that had experienced little besides persecution at the hands of the state, but that had gone on to create institutions, mores, and codes of conduct amenable to its members. Historian David Beito has frequently examined the role of private charitable associations in the era before the welfare state began offering compulsory charity to all. Economist Peter Leeson has examined the historical practices of pirates, gypsies, monastic communities, and others who often live far from the state's influence, to seek out new ways of understanding law and justice without any government as we might know it.

Some of these scholars are avowed libertarians; others are not. All of them belong to a relatively new way of thinking in the social sciences, one that does not begin by placing the state at the center. Collectively, they are part of a welcome new development. Possibly even a revolution. Historians are now writing and thinking about vast realms of the human experience that formerly might not have counted as "history" at all. In many of them, the state has not been an orderer, or a reformer, or even a ruler. It has been an unwelcome intruder, or shunned by all parties, or simply unavailable. Theorists of the state don't like to consider these cases, but they are real, and they are increasingly well-documented. They too are a part of the human story, and they suggest that the state's role can afford to be diminished, perhaps radically.

Political theory has lagged behind, in my view. Yes, it is called *political* theory, but to presume from this descriptor alone that all problems at hand require a political solution is to beg the question. The centrality of the state in our social thought may have driven to the edge some things that have an equal or greater claim to consideration as we look to solve difficult social problems. I certainly believe this to be so. To make the case properly, however, I will first need to present an alternate view of the state, one that draws on several of the dissenters we have already examined, and that questions what canonical political theory often takes for granted.

NOTES

1. Paine, *The Rights of Man*, Vol. 2, chapter III: of the old and new systems of government. In *The Writings of Thomas Paine* available at http://oll.libertyfund.org/title/344/17372.
2. There are many others whom I have neglected, because the history of liberalism is not the subject of this book. Perhaps another time I will have occasion to write about Montesquieu or Tocqueville, to name just two, but it suffices for the moment to say that liberals have long been in the minority, both across historical time and within the canon of writers whom we signal out as particularly important when studying political theory.
3. One need not reject formal logic to be a Hegelian or a Marxist, but Friedrich Engels certainly did. See Kołakowski, 320–321, for Engels'

views on the matter, as well as a very kind attempt at reconciling Marxism with formal logic, Engels' protests notwithstanding.

4. In part, this comes from the experience of reviewing unsolicited manuscripts for a major think tank. But for a published example, consider House's *Philip Dru: Administrator*.

5. Bellamy's *Looking Backward* was extraordinarily popular from the moment of its publication in 1878, despite its appalling economic ignorance. Bellamy boasted of having read the classical economists, but his work evinces absolutely no understanding of them.

6. F. A. Hayek, *The Counter-Revolution of Science*.

7. Popper, *The Open Society and Its Enemies.*, vol II: *Hegel and Marx*, pp. 298–299. Italics in the original.

CHAPTER 9

The State Is a Bundle

In Part I, we considered many different theories of what government is purportedly for. Throughout it, we set aside some thorny philosophical questions. It is now high time to address them. The first is that of *teleology*, or *final causality*. In Aristotelian metaphysics, the *final cause* of a thing is something like that thing's destined purpose. A final cause describes what a thing is *meant* to be. All things had a final cause, Aristotle believed.

Although to modern ears it may sound strange, the notion of the final cause remains with us to some degree, even if we don't typically use those words. Thus a non-scientist might say that an acorn is destined to become an oak, or that hands were made to grasp and manipulate things. Aristotle claimed that men were likewise made to take part in a *polis*, which is the only social context in which the goal of *eudaimonia* can be achieved.

Scientists are reticent about inferring purposes in the natural world, a move that seems to imply the existence of a consciousness that designed various natural features and parts of living organisms. Modern life sciences have found that they have no need of this hypothesis. As a result, Aristotle's ideas about teleology have long been set aside in the natural sciences. In the social sciences, this setting-aside is more difficult to do, insofar as every individual person clearly does have a consciousness and can express purposes for their actions. As we have seen, some have argued that institutions, nations, and even history itself have a similar purpose, and that we must strive to understand these supra-personal purposes if we are to understand the purpose of human society as a whole.[1]

© The Author(s) 2017
J. Kuznicki, *Technology and the End of Authority*,
DOI 10.1007/978-3-319-48692-5_9

Answering the question of what government is for thus brings us very close to questions of teleology: What is government's destiny? What is it developing toward? Like most moderns, modern historians generally don't like teleology. Even many philosophers have grown allergic to it. We are altogether comfortable describing the teleological views of others. But we shrink from proposing a teleology of our own. We are wary of saying that history has a necessary direction, or a destiny that it is trying to fulfill.

The past did not share our reticence. Authors as diverse as Plato, Machiavelli, and Hegel have all given accounts of government that are to a high degree teleological, and in each case, the teleology points to something outside the human individual. As we have seen, one of the remarkable things about these teleological accounts is that so many of them have assigned to the state a near-omnipotent role in ordering society. When we consider the *telos* of the state, individual lives must yield, so that the state may supply order.

We have also seen some dissenters. A few, like Augustine, characterize the worldly state's purpose as an evil one; a state's destiny may be to serve as an affliction, and to do so for as long as God wills it to be. Only a few thinkers, however, have questioned the role of teleology itself: What if the state *doesn't have a destiny*? What if its purposes change arbitrarily over time? What if the state only came to exist by some monstrous accident? (At times, I will grant, Rousseau seems to approach this view, although I certainly do not agree with him that private property is monstrous).

Post-Enlightenment philosophy has seen a sustained assault on the idea of natural or divinely ordained teleology. We can see this change taking place within the history of political theory as we recall some of the later authors we have considered: Marx believed that the state was contingent on other historical facts. The state was a means only to the ends of the ruling class, it depended on economic factors, and it would eventually disappear. Oppenheimer thought that the state had begun as an association of bandits; it won whatever legitimacy it accrued in the meantime by fending off the rest of the bandits. In Marx's case, the purpose of the state followed from the conditions of production. As such it can hardly be called a "destiny" at all; it is simply a consequence. For Oppenheimer, the state's purpose—if it has one—has been incoherent over time, because it began by doing one thing, and it ended by suppressing the very same thing.

The modern attack on teleology has been broad-based and widely sustained. Teleology's fall is also part of what Nietzsche meant when he declared that "God is dead": There is no divine orderer to the natural world, Nietzsche held, and we moderns are increasingly aware of it. There is no orderer to the moral world, either. And—of course—there is no orderer to give a lasting purpose to the state. If Nietzsche is right, then government exists only for the reasons that we supply along the way, and it may exist for no valid reasons at all.

So was Nietzsche right that God is dead? It's a big question, and it's not one I can answer here. But it does seem clear that governments can be re-purposed all too freely, and it might simply be that government, like many other things, doesn't have a built-in teleology. It is only as good as the purposes that individuals give to it. Difficulties with teleology do not mean that all government is necessarily illegitimate. But they do mean that many of government's alleged purposes are now doubtful, particularly those purposes that are purportedly greater or larger than the individuals who make up a given society. Connections to the divine, world spirits who guide history, and the exercise of martial virtue for its own sake are all purposes greater than those of individuals, and they are all to be doubted: A modern, post-teleological government should reject all of them.[2]

Viewed from the standpoint of *methodological individualism*, the state's teleology rapidly falls apart as well. Methodological individualism is an approach in the social sciences that holds that individuals are the only acting forces in society. When we speak of social classes, say methodological individualists, we are not speaking of acting agents at all. We are at best speaking of tendencies and clusters of behavior and ideas, and yet every social class in turn is made up of many individuals, each with their own wills and goals. They cannot be understood otherwise. And the same can be said of the state: It is not Germany that goes to war with France. It is the German government's leadership that has made the decision, and it is the individual German and French soldiers who fight. What if the state is likewise just made up of people, deploying the political means through a common institution?

To a methodological individualist, much of political theory amounts to various *asserted norms of behavior*. These norms are offered by one individual, the theorist, to other individuals who are his readers. The theorist typically asks them to treat some concerns, the state's concerns, as though they had greater priority than others, and he offers reasons for this priority. But upon examination, the methodological individualist will find that the state's concerns all turn out to be only the aggregated concerns of other individuals.

But if the state has no fixed goal, and if all goals are individual goals, what is the conceptual glue that holds it all together? Why do we even think of the state as *one* thing—particularly over long periods of time, during which the desires of individuals are subject to change? We must be careful not to sneak back to teleology here: We can't say that the state is the thing that *should* behave with the purposes that John Locke assigned to it. This places an earnest wish where a description ought to go.

Is there a way to describe the state that (1) clearly references actual entities and occurrences and (2) is not teleological or normative in character?

To answer, I will offer an extended analogy. David Hume argued as follows about the self:

> There are some philosophers, who imagine we are every moment intimately conscious of what we call our Self; that we feel its existence and its continuance in existence; and are certain, beyond the evidence of a demonstration, both of its perfect identity and simplicity...

> [F]rom what impression could this idea be derived? [...] It must be some one impression, that gives rise to every real idea. But self or person is not any one impression, but that to which our several impressions and ideas are supposed to have a reference. If any impression gives rise to the idea of self, that impression must continue invariably the same, through the whole course of our lives; since self is supposed to exist after that manner. But there is no impression constant and invariable. Pain and pleasure, grief and joy, passions and sensations succeed each other, and never all exist at the same time. It cannot, therefore, be from any of these impressions, or from any other, that the idea of self is derived; and consequently there is no such idea.[3]

In short, Hume challenged the idea that introspection supplied evidence for the existence of a unitary self. He claimed that when he introspected, he didn't find any one thing at all. Instead, he found only a bunch of discontinuous sensory experiences and impressions. All of them were variable. All of them came and went. But then where was the constant thing called the self?

States appear to occupy a place analogous to Hume's self, as described above: Being governed by a state is also a set of discontinuous experiences and events. All of us have experienced being governed; all of us have received specific commands and prohibitions. Yet one could conceivably remove any one of these commands or prohibitions, and the thing we call the state would remain. It would be a somewhat different state—it might be worse or better—but it would still be a state.

In other words, the state is a more or less organized bundle of uses of the political means. State agents are those who act in pursuit of the deployment of the political means to one end or another. State agents, tools, commands, and prohibitions are all capable of being removed and replaced. The responsibilities of citizens and functionaries may come and go. Still the state remains, at least as a thing that we speak and write about.

In fact, the apparent purposes in the state bundle sometimes sit quite badly with one another, even in a single time and place. The state both subsidizes corn syrup production *and* applies punitive taxes against its consumption. The state licenses alcohol production *and* discourages its consumption. The state incentivizes private sector employment *and* taxes private income, which to some degree dis-incentivizes private sector employment.

The bundle theory comes into even sharper relief when we consider states of different times and places. At some times, states have governed the price of bread, and at other times they have not. At times they have declared in favor of a state religion, and at other times they have refrained; at still other times, they have suppressed the practice of religion altogether. By various measures, there have been larger state bundles and smaller ones.

Consider the state bundle of seventeenth-century France. It contains all sorts of things that the state bundle of twenty-first-century America does not. In nearly all particulars, our state declines to regulate the religious beliefs of its subjects. This is quite unlike the Old Regime. Our state does not exact internal tariffs or compulsory labor. Our state declines to prosecute the former crimes of blasphemy, sodomy, and witchcraft. Our state is constitutionally forbidden from establishing a hereditary nobility, whereas the Old Regime made hereditary nobility a cornerstone of society.

Yet our state does many additional things that the seventeenth-century French state declined to do. In the recent history of the United States, compelling the purchase of health insurance has been added to the bundle. Not so long ago, compulsory racial segregation dropped out of the bundle, and compulsory racial integration replaced it. None of these were essential to the state-ness of the state, yet in their time all were very clearly state-related things.

This is not to reproach states for failing to adopt a uniform program across all time and space. It would be unreasonable to demand that. It is only to point out that the unity of the state seems more and more to vanish the longer we look at real-world states. Aristotle had held that the *polis* was properly defined by the constancy of its governing institutions; citizens may come and go, while the *polis*, which is the state, endures. We may now say that Aristotle was wrong. The *polis* does not endure, because the parts of it come and go as well, just like individual citizens.

It is sometimes claimed that the messiness of the democratic process is to blame for states' varying and conflicting purposes: Democracies lack the unitary nature of an individual will, and one possible corrective to democratic failure is to restore an element of unitary will to governance.[4] And yet, as we see from the histories of authoritarian regimes, democracy has no monopoly on incoherence. Tyrants contradict themselves too, and state bundles seem to be messy regardless of who compiles them, or how.

What if we began to remove items from the bundle, one by one? At what point would the bundle stop being a state? Suppose a state only adjudicated disputes over property, but it made no promise to repel foreign invasions? Or vice versa? Could a state only regulate ice cream sales, and absolutely nothing else, and would we still call it a state?

More seriously, what about states that have no fixed territory, or only a minuscule one? Such entities exist in our own world, and they have for centuries. The Sovereign Military Order of Malta is the world's oldest chivalric order. It is indeed legally sovereign; in diplomacy, it generally enjoys the same protections that a sovereign state could expect. It has a flag and a coat of arms. It is a permanent observer at the United Nations. It has its own currency and postage stamps. But it has not been sovereign over the island of Malta since 1798; its only territory is of an extent comparable to those of conventional states' embassies. It is a humanitarian organization affiliated with the Catholic Church that does relief work worldwide. It helps the poor and disabled. But it barely governs at all.

We have just encountered what philosophers call a *sorites paradox*, or a problem of heaps: Everyone knows what a heap of sand is. Everyone can identify heaps of sand when they see them. But what is the smallest number of grains of sand that can make up a heap? If we take a heap and remove the grains from it one by one, what essential change causes it to go from heap to not-heap? Are three grains a heap? Is one grain a heap? Can a heap have zero grains? This last possibility is not so easily dismissed when we recall that in mathematics the empty set remains a set in good standing. If "a heap" means something like "a set," then an empty heap isn't an absurdity at all.

Applied to the theory of the state, the sorites paradox can yield surprising results. It could be asserted that the state that regulates ice cream sales alone still counts as a state, exactly as some might say that a single grain of sand is nonetheless a heap. Or consider Underwriters Laboratories, which regulates electrical equipment with the blessing of the U.S. federal government. Seemingly it approaches being a state-within-a-state, a government that does only one thing and nothing else. From there, we might be forced to admit that we are surrounded by tiny, improbable statelings: Mall cops, private arbitration firms, gated communities, kosher certification services, and cryptocurrencies all do at least one of the things that states have sometimes (or often) done, albeit with the approval of a higher, supra-state, without which their work could be of doubtful effect. The theoretical private defense agencies of anarcho-capitalism begin to look suspiciously like states as well; they too are small-ish bundles of the organized use of the political means.

Now consider the case of the zero-grain heap. Can it really exist? If a heap *can* have zero grains, then perhaps a state can exist that does no governing whatsoever. It might even have no capacity for action. This at first seems to suggest an ordered anarchy—but is it really proper to call an anarchy a "state"? The two are at least conventionally understood as opposites. A properly ordered anarchy might entirely lack the use of the political means; its "state" for want of a better word would be a zero-grain heap. When we consider that the etymological ancestors of the word "state" originally referenced a *status,* this is perhaps not so absurd as it seems.

To be absolutely clear, we have not yet seen such a society. If we identify the state with a bundled deployment of Oppenheimer's political means, then we must conclude that the real world's so-called anarchies have been nothing of the kind. Places like Somalia are not properly called anarchies because they are rife with the political means, perpetrated by many different organizations. As a result, they should instead be thought of as *governed societies,* albeit societies that are governed in an exceptionally chaotic and unjust manner, by many small state-like entities. At the risk of paradox, these anarchies are governed far more than they need to be.

Is the degree of *organization* the better dividing line between statehood properly speaking and this type of degenerate anarchy? Is it possible that the only legitimacy the state can be said to have comes insofar as it makes predation relatively regular, as opposed to bandits, whose predation is irregular, but whose actions remain in a sense state-like? If so, as Mancur

Olson noted, we may have reason to prefer a unitary state, which consists of stationary bandits, over a situation in which we were at the mercy of roving bandits. And this would be true even if we found no other reason to consider a unitary state legitimate.

If we were to eliminate all of the state's aspects, rather than merely delegating the political means to various irregulars, could we at last behold an empty bundle? And might we discover that we preferred the stateless society that remained? These questions are hard to answer in part for lack of data. Contemporary philosopher Anthony de Jasay gives much credence to the conquest theory of state formation, exactly as Oppenheimer did. De Jasay wrote as follows in one of his own books, which is also titled *The State*:

> If ... as a matter of repeated historical fact it took violence to impose a state upon [people in the state of nature], it seems pertinent to ask, Why does standard political theory regard it as a basic verity that they preferred the state? ... Do people in the state of nature prefer it to the state? and Do people, once in the state, prefer the state of nature to it? These questions very sensibly allow for people's preferences to be related, in some way, to the political environment in which they actually happen to live. However, once they are framed in this way, they are seen to have a peculiar character. ... People who live in states have as a rule never experienced the state of nature and vice versa, and have no practical possibility of moving from the one to the other ... On what grounds, then, do people form hypotheses about the relative merits of state and state of nature?[5]

We don't have anyone's revealed preferences about the choice between anarchy and state. Even avowed anarchists, who say that as a matter of principle they would prefer to live in an anarchy, are in today's world restricted merely to talking about it. They can declare their preference, but they cannot reveal it through action (Violent anarchists, mind you, may reveal their preference for violence, but they do not reveal a preference for living in anarchy. They are incapable of doing so, because no anarchy exists for them to live in).

Each of us has only the bundle of the state, or, at best, one of several possible state bundles to choose from, if we are capable of relocating. The state bundle's contents have changed over time, and we may change which bundle we experience by moving, but we are unable to choose an empty bundle. Even medieval cities, discussed briefly in part one, were founded, qua states, as a defense against other very state-like entities; as a result, even they do not evince a clear preference for statedness over anarchy. One might even argue that insofar as the choice to form a city and seek protection was compelled, the more authentic choice of the burghers might have been not to form a city government at all, but simply to live in an ungoverned urban area.

Some may argue that the omnipresence of the state is strong evidence that the alternatives are unworkable: If the alternatives were—or are—workable, why don't we see them? Those who argue in this way are correct, I believe, with two caveats. First, strong evidence does not constitute proof. And second, the alternatives may only be unworkable for us—and not for all time, or for all people. That would be a claim unwarranted by the evidence. The need for defense against hostile states may itself be sufficient reason for having a state, but this does not mean that states are justified. It may only mean that we are stuck in a bad local equilibrium.

These questions can't be answered. But some questions can: If we accept that the state is a bundle, and if we accept that its contents have changed tremendously over time, we find that the elements of the state are united not by a purpose, but by a *method*, which is no more and no less than the political means; and by a *convention*, namely the political philosophers' habit of considering many different forms of organization of the political means (but not all such forms) under a common rubric.

If it appears uninspiring, I will plead guilty as charged: I do not believe the state should be inspiring. As already discussed, I happen to believe it should be much smaller than it is today, but I am even more sure that it should occupy a much less exalted place in our mental lives. To resort to government in the forms commonly spoken of as the state, and falsely praised to the heavens over the centuries, is to resort to a very blunt and very ugly instrument. We have no reason to be especially proud of it. We have good reason to be wary.

It should also be stressed that this description of the state does not justify the state's continued existence. Nor can it be made to do so: Simply being incapable of finding a non-state solution to a given problem does not mean that no such solution exists, and that the state must therefore be justified. An argument of this form, which I do not advance, would be congruent to one that is already well known, and known to be logically flawed, namely the so-called God of the gaps argument, which both theists and atheists alike have condemned: If we find the proof of God's existence *only* in the things that science cannot explain, then we have no adequate proof of God's existence at all. Likewise, if the best that philosophical defenders of the state can do is to advance an argument from ignorance, then they have no adequate justification for the state. Other justifications may exist, but this one will not do.

Of course, there are other problems with my definition as well. Three of them are as follows: First, is my definition complete? What about other commonly advanced attributes of the state, including legitimacy? Second, if we accept the definition as complete, who does the perceiving, and how do they do it? And third, what about the moral status of unrequited transfer—in accepting this definition, are we committed to a normative rejection of the state? I will consider each of these in turn.

NOTES

1. God, of course, may be said to supply the consciousness that at first seems requisite in teleological accounts of both nature and history. This move raises the problem of theodicy we encountered discussing Hegel. But it may also be possible to have teleology *without* consciousness; if so, then we have no need of the God hypothesis. Marx's stadial account of social development might be called teleological in this way.
2. A particularly daring advance in this direction comes from Irwin, *The Free Market Existentialist*, which begins with Nietzsche's claim, progresses to the existentialism of Jean-Paul Sartre, and ends, remarkably, in free-market capitalism, arguing persuasively along the way that existentialists were wrong to embrace Marxism.
3. David Hume, *A Treatise of Human Nature*, Book I section iv: Of personal identity. http://oll.libertyfund.org/title/342/55127.
4. Then again, if Hume is right—and I think he is—individual wills are not unitary either. They too are disjointed and incoherent. ("I will lose weight … I will eat this entire tub of ice cream. … I will lose weight …")
5. de Jasay, *The State*, 1.1.6. http://www.econlib.org/library/LFBooks/Jasay/jsyStt1.html#1.1

CHAPTER 10

Some Objections to the Theory

Is this definition compatible with Max Weber's legitimist approach? Other accounts of the state certainly exist; the best known, of course, is Max Weber's. In Weber's account, the state is the entity that successfully claims a monopoly on the use of legitimate physical force within a given territory.

Weber's definition correctly identifies *modern* states, or at least it identifies what modern states aspire to be, but it clearly fails to include premodern states, as Weber himself admitted. Legitimacy in the eyes of barbarians mattered little to the Roman Empire as long as the conquered provinces paid tribute; legitimacy certainly mattered at the center, but even there, it mattered primarily among the citizens, and the very question of legitimacy to the enslaved members of the population would have been difficult to pose seriously. Medieval and early modern polities meanwhile had complex overlapping structures of authority, in which any claim on territorial monopoly was unlikely to go beyond words on a page, if it was even asserted there.

The thing talked about in political philosophy, that which has been a constant subject of discourse from Plato to the present, is *not* the same thing as Weber's state. My definition tends rather to include these earlier models. It also tends to acknowledge a significant gray area that can never be entirely eliminated, within which some entities are state-like to varying degrees. This Weber's definition cannot do.

Now, if the conquest theory of state formation is generally true, we should sometimes expect to see competing state-like entities in the same territory, each of which currently has no monopoly, and each of which has

© The Author(s) 2017

J. Kuznicki, *Technology and the End of Authority*,
DOI 10.1007/978-3-319-48692-5_10

at least some claim to legitimacy. And indeed we do. Often states coexist in a state of conflict or tension for long periods of time, their claims to monopoly notwithstanding. I would suggest that they remain, nonetheless, states, in that they can be profitably analyzed as organized deployments of the political means, about which much political theory remains plausibly applicable. In the American colonial era, the northwest frontier was home to many different states, each of varying strengths over time— British, French, and various Native American polities all made claims to the territory, and for much of this time, no clear monopoly existed.[1] Loyalties could be found all around, sometimes to one or another state, and sometimes to more than one state simultaneously. Legitimacy was contested and fluid. None of this meant that the territory was ungoverned, or that no states existed there.

In the present day, organizations like ISIS may be described as aspirant states in Weber's sense, in that they hope one day to command a legitimized monopoly. Yet even now, they do many of the same things that recognized states commonly do, including punishing crimes, adjudicating disputes, and offering transfers to the needy. It is difficult to say exactly when organizations like these finally become states, particularly as the monopoly status of *any* state is never entirely complete. Even the U.S. federal government in a sense competes with criminal gangs, which also offer a set of services, also in response to perceived needs, and also through unrequited transfer.[2] Whether the gangs are *legitimate* or not would appear to depend on whom one asked, and in what circumstances.[3]

Weber's definition also seems to rest on a definition of physical force that may or may not be obvious or unproblematic to readers. In my own dealings with the U.S. government, for example, I have yet to experience a single act of direct violence. There have been implicit threats, true, but nothing more. But one thing clearly has taken place, namely unrequited transfer, both to and away from me. This state of affairs probably holds true for most Americans, suggesting that unrequited transfer describes state action more fully than the use of force or implicit force in most cases.

Finally, and in any case, the prohibition of some actions and the compulsion of others—often taken to be defining actions of the state—will inevitably require resources to be carried out; although these are both ubiquitous parts of the state bundle, they both require an initial appropriation of resources if they are to take place at all. Unrequited transfer is anterior to them both.

None of this is to say that Weber's approach to the state is without its use. In pointing up the aspirations to legitimacy and to territorial monopoly, Weber has identified two things that set modern states apart from earlier precursors. But this need not prohibit us from deploying the word state ourselves, in a somewhat different manner, and with appropriate cautions.

Who decides when something has become a problem appropriate for state intervention? One strength of my definition of the state, I think, is that we do not have to decide for ourselves when state action is justified. It makes no necessary normative claims, and yet it is potentially compatible with many of them. It is significantly value-free.

Obviously various governments will have diverse apparatuses to decide when intervention is appropriate. In all of them, we must admit that there have been frequent false positives: It is hard nowadays for us to see the presence of interracial marriage, or of religious diversity, or the perceived presence of witchcraft, as problems in any sense whatsoever—much less as problems needing a state solution. But they have been deemed such in the past.

A pervasive conflict of interest likely exists here, in that state actors themselves are typically the ones who choose whether a phenomenon constitutes a problem requiring their own action. *Of course* they will be likely to hold themselves competent to make such decisions. And democratic methods of decision making are not clearly better here than autocratic ones; even democracies have shown a tendency toward false positives. Democracies also introduce well-known principal-agent problems: We may well vote for the representatives who decide which things go into the bundle, but we do not necessarily approve of their choices, and removing an item from the bundle is frequently much more difficult than introducing one.

Libertarianism stands strong when it notes the many times that the state has manifestly failed to solve the problems that it has arrogated to its bundle. Libertarianism stands even stronger when state solutions achieve perverse outcomes, as is often the case, or when it becomes clear that certain items in the bundle have little or no business being there. But there are times, however, when libertarianism may overreach, as will be seen in considering our next objection.

Is the state necessarily illegitimate? Observing that certain actors have identified a problem and placed it in the state bundle does not make a normative judgment. But a very simple additional step—declaring that all unrequited transfers are illegitimate—does make a normative judgment, and a sweeping one at that.

This line of thought is often advanced by the political theory known as *deontological libertarianism*. Deontological ethical theories give no regard to the consequences of an action but ask only about those considerations taken to be necessarily true before, during, and after the action in question. Deontological libertarianism's key argument runs something like this: For all property that is held legitimately, unrequited transfer is always morally wrong.[4] Whenever transfers of legitimately held property occur, they must happen only by the free consent of all parties. If not, the transfers are illegitimate. But if states *always* perform unrequited transfers of legitimately held property, then states are always illegitimate.

Many find this form of libertarianism preposterous. It sometimes seems to prove far too much,[5] and it may appear to commit the arguer to positions that are exceedingly hard to defend.[6] Yet the argument remains capable of haunting even the most well-regarded of unrequited transfer proposals: Could we think of *no* other ways to solve the problem at hand? Is this really the best we can do?

One can typically think of at least a few possibilities, and if not, libertarians will be delighted to point some out. But why then does our society not deploy these other methods? A deontological libertarian would hold that *any* state of affairs that did not entail unrequited transfer (or other deliberately perpetrated moral evils) would *always* be preferable to any state of affairs that did. Let the consequences fall where they may, says the deontologist. We shall not take part in evil.

Not everyone wishes to be so insensitive to consequence, of course, and the typical response to deontological libertarianism involves pointing at the good consequences of state action: Tolerate just a little unrequited transfer, and thousands will be fed and clothed, perhaps. This alone is more than enough to settle the argument for many: We break this one little ethical rule, and we realize a great deal of good in the process. So much the worse for the rule. Almost invariably, thought experiments can be constructed that are capable of nudging a deontologist into a consideration of consequences.

But thought experiments alone are not sufficient to construct a rationale for action in the real world. Do our real-world public policies have a similar moral legitimacy? We already know how a strict deontologist would answer. A weaker form of libertarianism, but I think a much more defensible one, would say that one or more practices *might* exist that (1) would *not* entail unrequited transfer and (2) might achieve equal or better results in solving the problem at hand, subject to empirical verification. Any solution that possessed both of these attributes would obviously be preferable to any evaluator, provided only that they have a nonzero weighting to the injunction against unrequited transfer.

This observation would hold even for the suppression of the most heinous crimes: if there were a system of purely voluntary murder deterrence that deterred murder as effectively or more effectively than the current, state-run system, we would be obliged to adopt it as long as we gave some weight, however small, to the injunction against unrequited transfer. It is only by assigning a positive ethical value to the act of unrequited transfer itself, and by further asserting that positive values such as this one may in some way cancel harms, that we might reach the opposite conclusion. Both assertions seem implausible.

Indeed, for all policies that satisfied (1) and (2), we would not even have to specify how heavily we weighted either the injunction against unrequited transfer *or* the injunction to consider consequences. We would still know, regardless of our weighting of these two moral systems, that we ought to prefer the voluntary means over those of the state.

Further, at least some otherwise second best voluntary solutions from a purely consequentialist standpoint may also become preferable to state solutions whenever we accept a nonzero weighting of the moral injunction against unrequited transfer. Their number and extent would vary with the weighting of the two injunctions in our moral calculus, in ways that we do not need to explore here, but that perhaps should be explored elsewhere: It would seem obvious, to me at least, that the vast majority of moralizing in politics proceeds from either strict deontology or strict consequentialism, perhaps narrowly applied so as to seem more plausible in the moment. But moral reasoning in everyday life is not so dogmatic; if it were, it would frequently turn absurd.

In the real world, we happily assign weights to conflicting moral intuitions. Virtually no one is purely a consequentialist or purely a deontologist, and again, thought experiments readily establish the point. Would we be justified, for example, in taxing a randomly selected dozen or so Americans at a very high rate—perhaps a flat 90% of their total income— and transferring the proceeds directly to poor Africans, who endure a vastly lower standard of living? From a purely consequentialist standpoint, one might argue that, relatively speaking, the Americans can do without the extra income, that the tax affects only a few, and that the transfer has the potential to alleviate poverty among perhaps hundreds of people. Yet this is a solution that no one has ever proposed in the real world, for reasons that are most easily described as answering to the deontological objection. Note that our difficulties with the scheme disappear entirely when we imagine a dozen or so Americans *voluntarily* living in extreme simplicity and donating a large share of their income to charities helping Africans. We may even find their actions praiseworthy.[7]

Powerful intuitions are at work here, and powerful intuitions will often conflict, even within one person (this, remember, is only to be expected if David Hume was correct about the nature of the self as a bundle). The intuition against unrequited transfer is strong and broadly shared; sound or not, this intuition is one of the most effective selling points in the libertarian repertoire. But there exists as well an undeniably strong intuition that taxation may *sometimes* be justified. Referring to widely shared intuitions alone may leave us at loggerheads, particularly when we consider the problem in the context not simply of an individual mind, but of a society, in which various people will find either of these intuitions persuasive to varying degrees.

The intuition that supports unrequited transfer may be correct in some cases, but it will not do, as some have done, to gesture grandly at "civilization" and declare that taxes are the price we pay for it. If that gesture alone sufficed, taxation would know no limits. The argument for taxation can only be as strong—at best—as the efficacy that a specific state action has in achieving a specific good purpose, minus any moral ills of taxation and the deadweight loss inherent in taxation itself. Meanwhile the intuition against taxation is accepted by nearly everyone, at least part of the time, in that most people appear to believe that they hold the remainder of their property justly, rather than being obliged to turn it over to the state, which they ought to do if they held it unjustly. Each dollar that remains in a taxpayer's pocket is to some degree a rebuke to the ones that are taxed away.

We may therefore be justified in placing *some* weight on the objection to unrequited transfer, without declaring that it immediately robs the state of all legitimacy. This stance seems to square with most people's common sense: Unrequited transfer of property is perhaps one evil among many to be combated, and its weight need not be decisive in all cases. But to what degree must it be decisive, given our weightings of various intuitions? We may not yet know, and this ignorance ought to trouble us more than it seemingly does.

Taxation can be theft, as libertarians often say, and yet there can be worse things in the world than theft, and we can perhaps use theft to stop them. The more incisive question is whether any better methods are available. Few, whether libertarian or otherwise, have been willing to seek out these other methods: The deontological libertarian is content to repeat endlessly that taxation is theft; the modern liberal is content to repeat, also endlessly, that taxation is in some way the thing that makes us civilized. The result is a sterile impasse, not a search for better solutions.

It should rather be incumbent upon policymakers, theorists, and also ordinary citizens to explore non-state solutions with at least as much energy as we now devote to promoting or condemning state solutions, if not more. The answer can't possibly *always* be yet another government program; our choices do not seem appropriately limited by recourse only to unrequited transfer, or to doing nothing. Might we eventually think of better methods than government at remedying a problem at hand? Methods that require less, or no, unrequited transfer? Methods that realize better consequences? Or at least methods that realize consequences that are to our minds more tolerable than violation of the rule against unrequited transfer, however we choose to weight it?

Many such methods, though not all of them, fall under the rubric of market exchange. Many others are encompassed in those forms of civil society that are not, or are only partially, marketized. As Alexis de Tocqueville observed of early America, "I found... some kinds of associations of which, I confess, I had not even the idea, and I often admired the infinite art with which the inhabitants of the United States succeeded in setting a common goal for the efforts of a great number of men, and in making them march freely toward it ... there is scarcely so small an enterprise for which the Americans do not unite."[8] In attempting to solve difficult social problems, the state should be but one choice among many, and the methods of private association should always be carefully considered before we resort to the state. It would make us all rather embarrassed, I should think, if we had been reaching for the political means when other means were lying close at hand, and when these means would realize equal or better consequences.

An *assessment* of state methods now becomes important as well. And many problems lurk here. It is often relatively easy to assess whether a government program has had a positive effect within its declared scope of operations. For exactly this reason, states may even incline toward measures whose good effects are easy to see, and whose bad effects are well-hidden.[9]

Assessing a measure's full effects, including accounting for hard-to-see considerations like the deadweight loss of taxation, the changed incentive structure of society, and other unintended consequences, is usually much more demanding. State agents understandably prefer it this way, making a full and accurate account of any particular policy difficult. To the extent that biases exist for these reasons, they will incline us to state-based solutions.

Comparing a government program to the absence of the same government program presents further difficulties. It does not suffice to consider how things stood before the program was implemented: There is in fact a universe of possible societies that all lack the program, and that either compensate for its absence in various ways, or simply ignore the problem and do nothing about it directly. The society that came immediately before the program at hand is only one such possibility. Worse, it presumably wasn't doing such a good job of dealing with the problem (or, perhaps, of assessing it appropriately) at the time of the program's implementation. Had that society been doing acceptably, the citizens and/or their agents in the government might not have acted to solve the problem through state-based methods. But other, wholly private setups may do better than either the one that they implemented or than the status quo ante. One need not be a deontologist to suspect as much. All one needs to do is recognize that we do not yet know all that we might like about the vast field of public policy.

In part as a result of the authoritarian bias that I detect in the western canon, I believe that philosophers of all sorts have been neglectful of the vast design space that confronts them. Choosing *any* political system, or any social arrangement at all, whether normatively or practically, will always mean neglecting an enormous number of alternatives, most of which probably have not yet been imagined, let alone examined or defended. In light of this neglect, much of the argument in the field may amount to affirming the consequent: If a political system is the best, it will have benefits x; the political system I name has benefits x, and therefore it is the best. This will not do. Other systems may have equivalent benefits beyond the systems we are capable of naming. They may even supply the named benefits at lower costs or to a greater degree.[10] Thus, Rawls commits an error when he writes that, for either individual actions or for social institutions:

> [T]here is no way, let us assume, for the parties to examine all the possible principles that might be proposed. The many possibilities are not clearly defined and among them there may be no best choice. To avoid these dif-

ficulties I suppose, as before, that the choice is to be made from a short list of traditional and familiar principles.[11]

This is the very assumption that we should not be making. Our choice is not actually limited to a short list of traditional and familiar principles. As a result, deontological libertarianism places the state under *some* indictment, but we may have great difficulty saying just how severe the indictment is. We do not, after all, know the full list of alternatives; and even if we did, we would still need to evaluate them all according to the weightings we assigned (1) to the injunction against unrequited transfer and (2) to the consequences we believed would follow from each choice.

Deontological libertarianism thus makes a precipitous move in claiming that its indictment of the state is decisive all by itself, and that the state must be demolished come what may. But, although precipitous, it could turn out in the fullness of time to be a *correct* move—and better, one that can be understood as correct not merely by today's deontologists, but perhaps by everyone who seriously considered the problem, consequentialists included. We just can't prove it yet.

Deontological libertarianism also faces another problem: it may be an example of a system that is justified only with reference to what Derek Parfit has termed its *ideal act theory*. Ideal act theories are those parts of a (presumably) larger moral edifice that articulate which principles would produce excellent consequences—if and only if all people always adopted them and always complied with them successfully. As Parfit argues, a full moral theory must contain much more than simply a set of claims about ideal acts. It must also account for failures of knowledge and will, as well as for earnest disagreements.[12]

Parfit did not ever intend ideal act theories to stand alone; for him, they are only one component of a fully developed moral system. But it seems clear that for many people, ideal act theories *do* stand alone. That is, they begin by envisioning the universal adoption of particular moral principles, with universal success in compliance. With these conditions as their ethical starting point, they move directly to claims about what states and other actors should do in the real world, disregarding the fact that the principles are neither universally held nor always complied with successfully by those who profess them.

This move seems commonest when we consider the vices of others. For example, an abstinence-only approach to sex education is an example of a bare ideal act theory, at least as it is commonly defended. It is rather obviously true that if all people complied successfully with an ethic of abstinence, then vastly fewer sexually transmitted diseases and unwanted pregnancies would result. This reasoning though does not proceed from an accurate depiction of how individuals are likely to act in the real world. It considers only ideal acts, demands perfect compliance, and disregards the consequences of anything less.

It's easy to imagine how deontological libertarianism might be similarly limited in its practical application: If everyone were always respectful of persons and of an acceptable rule governing the disposition of property, and if everyone always unfailingly honored contracts, and if all property transfers were likewise always voluntary, it's not too much of a stretch to suppose that this world might be an excellent one, and perhaps even a utopia, particularly if libertarian claims about the productivity of a pure laissez-faire capitalist society are also true.

But an excellent showing by ideal act theory is not enough to determine a system's suitability for the real world. The existence of a significant number of noncompliers who earnestly desire to set up a state (and perhaps a predatory one) would almost certainly mean that we never approached the situation just described. If the noncompliers banded together, they might even become the most powerful association of organized force in the society. They would then cheerfully begin the unrequited transfers that they had wanted all along. In short, a predatory government could all too easily arise. Recourse to the political means is after all not exclusive to state agents; all able-bodied human beings may deploy it, and the instantiation of a society that did not somewhere manifest the political means would necessarily require substantial discipline of every individual will.

Morality aside, extracting wealth by force is lucrative, meaning that the first people to try it will soon have much greater resources with which to consolidate their position. They can then use their first-mover advantage to set up a permanent state apparatus of whatever type they desire. The track record of existing states would suggest that it won't be a good one.

Someone employing a deontological approach might answer this type of objection by saying that the predatory government would at least be none of *his* doing, and that there is a moral difference between committing an act of aggression and allowing another to commit a similar act, particularly when trying to prevent that act would come at serious personal risk. He could with considerable justice protest that his hands were clean.

Yet if we are at all sensitive to consequences, that is, if we are even slightly committed to judging an action not merely in deontological terms, but by whether it leaves the world better or worse, we may not find this nonchalance very satisfying. I certainly do not. It may fall to us, then, to defend a minimal state as a protection against the threat of a maximal one. In simple deontological terms, we will have committed a fault. Yet we will also have avoided a catastrophe in the only way that we know how. In a world where not everyone shares his moral intuitions, the deontologist may find that he has no choices that answer tolerably well to his maxim.

For all of these reasons, the deontological argument against unrequited transfer seems less than decisive to me. It has considerable weight, but it can be defeated. In adopting the view that the state is an agency that solves perceived problems through unrequited transfer, we have raised the possibility that states are illegitimate, but we have not proven it. And states may even act rightly, at least to the best of our knowledge.

In short, if it were possible in all cases to achieve better consequences through private means than through state ones, then the weight we gave to the injunction against unrequited transfer would be irrelevant. We would prefer the private means. If equal or better results were possible in only *some* cases, then much would depend on how heavily one weighted the injunction, relative to the consequences to be realized.

Notes

1. See White, *The Middle Ground.*
2. See Fiorentini, *The Economics of Organised Crime.*
3. I would not allege that these entities are moral equivalents; the legitimacy in question pertains only to the possession of a monopoly on force, not on that force being put to normatively good purposes. In this context among others, the question of acquiescence under duress raises serious problems for Weber's theory of legitimacy.
4. A sentence like this one is incomplete *as an account of property.* I presume that it is possible to hold property legitimately in the initial instance, via means that are not articulated here. Some do not share this assumption, of course, and they at times have urged the general illegitimacy of property as a reason to support property transfers by the state. Arguments of this type fail immediately, in that the state would appear unable to confer legitimacy either: transfer logically entails ownership, and ownership itself has already been denied. As a result, the recipients in this scheme would appear

to have no better title than the former claimants; they too are unable to hold property legitimately.
5. Zwolinski, "Libertarianism and Pollution."
6. Rothbard, *The Ethics of Liberty*, part II, ch 14, argued that adults have no enforceable positive obligation to feed their children: Governments were not justified in performing the unrequited transfer needed to rectify the situation. As a result, parents could starve their children, and bystanders would have no right to interfere. The flaw in this position is elementary. Rights theory is founded on generalizations about adults. The theory therefore is grossly misapplied in this case. We could with as much justification apply rights theory to planets or pieces of music as we could to infants. For a sensible dissent, see Walter Block, "Libertarianism, Positive Obligations and Property Abandonment: Children's Rights," which explains how, even within a strictly Rothbardian framework, Rothbard erred on this question.
7. The propertarian solution to poverty in the developing world is all but certainly the correct one: Institutions and legal regimes must be developed that allow the poor to win and keep the proceeds of their labor, and that protect them against predatory corporations and governments, who are now all too accustomed to performing unrequited transfer. It is the lack of such institutions that has long kept the developing world poor, and not lack of generosity on the part of Americans. The classic study on this theme is de Soto's *The Mystery of Capital*.
8. de Tocqueville, *Democracy in America*, vol II, p. 897.
9. A particularly dramatic example of this phenomenon comes in the regulation of prescription drugs, where the failure to approve a beneficial drug is all but invisible; failure to approve is identical to the status quo, and thus it bears fewer costs for government actors themselves. See Isakov et al., "Is the FDA Too Conservative or Too Aggressive?"
10. One such system may be futarchy, as proposed by Robin Hanson: http://mason.gmu.edu/~rhanson/futarchy.html. Futarchy is amenable to a wide variety of modifications, none of them have been explored in depth, and it is just one possibility in the design space I reference.
11. Rawls, *A Theory of Justice*, p. 294.
12. Parfit, *Reasons and Persons*, p. 100.

CHAPTER 11

The Falsification of State Action

Much depends, clearly, on what we are capable of knowing about the science of governance. A central contention of this book has been that political philosophy has been overconfident about what it believes it can justify. Now, if I were simply looking at the biases of intellectuals in the course of western history and saying "we should consider moving *in the other direction*," my claim might sound like little more than a hunch. But that hunch is well supported by an argument made in formal and logical terms.

Let's start with another insight from Anthony de Jasay. De Jasay has argued that epistemology itself establishes what he terms the *presumption of liberty*.[1] His argument runs as follows.

There are only two types of substantive claims about the truth value of a statement: One may claim that the statement is true (or likely true), or one may claim that the statement is false (or likely false). Offering evidence that the statement is true is known as *verification*. Offering evidence that the statement is false is called *falsification*. The choice between verification and falsification is primarily one of efficiency, de Jasay writes. From one case to the next, one of the two choices is apt to be much easier than the other, and thus to offer more fruitful avenues of inquiry.

Now consider statements of the form "X should be forbidden from Y," where X is a set of persons and Y is an action. For this type of statement, verification and falsification meet very different fates. Someone wishing to *verify* the statement "X should be forbidden from Y" might be able to do so in a number of different ways, perhaps with reference to the direct harms of Y, or to the way that Y violates some agreed-upon moral norm,

© The Author(s) 2017
J. Kuznicki, *Technology and the End of Authority*,
DOI 10.1007/978-3-319-48692-5_11

or to the way that Y makes future coordination of social action more difficult. After that, we would have a *verified* reason to forbid X from Y, and the prohibition might be allowed to proceed.

But someone who wished to *falsify* the statement "X should be forbidden from Y" would have to defeat every possible reason that might ever forbid Y. That's literally an infinite set of reasons. The set includes a vast number of reasons whose disproof is trivial, but then, if we really do require a demonstration that the statement is false, then all of this infinite amount of trivial work must nonetheless be done. The set of possible reasons also includes many reasons that we have not even imagined yet, and if we fail to imagine them, then we cannot rebut them, and we cannot have met the burden of proof needed for falsification.

Taken together, these considerations guarantee in advance that no argument for any liberty—for any abeyance of a prohibition—will ever succeed. Yet this is profoundly implausible.

To avoid this absurdity, we must place the burden of proof on those who can more effectively bear it: We must oblige the statement's proponents to verify that "X should be forbidden from Y." Attempts to verify such statements (1) are at least potentially feasible and (2) are not evocative of a dystopia in which all imaginable acts are always forbidden. (By whom?) Placing the burden of proof on those who would enact a specific prohibition yields a fair fight, one entailing at worst only a limited list of prohibitions, all of which are set up against the background presumption that if an act is not clearly prohibited, it may still be performed.

Throughout part one of this study, we have seen various authors in effect try to meet the burden of proof for a surprisingly large array of specific prohibitions (and specific commands, which hold nearly the same epistemological status). Yet being able to meet the burden of proof to a degree that warrants action does not necessarily mean that the truth of the matter is settled forever, as it might be for a mathematical theorem. We may still act to prohibit, or to command, in the presence of a degree of residual uncertainty.

And indeed we do: the residual uncertainty cannot be avoided, because we deal here in general theories that are justified by means of singular empirical claims. This is the same class of theories about which the philosopher Karl Popper has argued, "[I]nference to theories, from singular statements which are 'verified by experience' (whatever that may mean), is logically inadmissible. Theories are .... *never* empirically verifiable."[2]

In Popper's admittedly extreme view, the verification of empirical claims is *nonsense*. But one need not follow him nearly so far in order to agree, more modestly, that a general, empirical theory that appears to have been verified by singular results could still be falsified later. It is in this epistemological category that all claims to the legitimacy of state action must fall.

In the context of state action, and of the claim that "X should be forbidden from Y," it is easy to see that X is typically not a specific person, but it is usually rather a general class of persons, including perhaps all persons entirely. It also generally includes people about whom the legislator knows very little. Even when the class consists of only a single person—Mr. Smith should be forbidden from leaving prison cell 304, perhaps—the command treats of a person whose circumstances are not fully known, and whose future can only be guessed at. The assertion that person(s) X should be forbidden, on the basis of known facts, from doing Y, therefore has the character of a *theory*: it makes a testable and universal claim about a class, not all of which has been or can be observed.

This means that the defensibility of a prohibition is always *provisional*, exactly as the theories of natural science are always provisional. Whereas type of evidence that falsifies a scientific proposition consists, as Popper argues, of a prediction that the theory makes that does not come true, the evidence that would falsify a prohibition could take various forms: We might discover that the supposed harm of the action is not a harm at all. We might learn that while the act is harmful, the prohibition itself generates harms in excess of the thing prohibited. We might reject, as many have done, the category of consensual harms altogether. We might discover wholly new and consensual ways of mitigating whatever harm prompted the prohibition. In each case, we will find that the justification for state action, which had heretofore only been verified, will now have been falsified.

Some may object that permitted liberties at least appear to be likewise subject to falsifiability: For example, where formerly the emission of mercury from coal burning power plants was considered irrelevant, and therefore permitted, science has since found significant harms that result from the practice. Prohibitions are therefore clearly justified in light of new data.

All of this is certainly true; but it does not constitute the falsification of an argument for a liberty. Recall that liberties—understood as specific exemptions from prohibition—cannot properly be argued for at all; it is only prohibitions that require an assertion and a defense. A situation in which a prohibition *should* be asserted, and in which it is not, does not constitute an exception to the theory. It is only a failure of a particular instantiation of the theory, and one that should clearly be remedied.

Like the theories of natural science, the theories of *political* science may be more or less difficult to falsify, of course, and they may be more or less deeply ingrained. I would not deny, on the one hand, that murder should be prohibited, or, on the other, that the available evidence clearly supports the idea that the origin of species lies in a process of genetic mutation and unguided natural selection. Both of these propositions seem vastly more probable than the alternatives. And yet it remains possible to imagine, for each, a body of radically new evidence or circumstances that could lead us to revisit our conclusions. In biology, we might discover that genes in fact track some other, as yet undiscovered process, one that explains both their operation and the (apparent) outcomes of natural selection. In law, we might consider revisiting the prohibition on murder, if, for example, futuristic technology made it possible to generate moment-to-moment backups of a person's complete mental and biological state, and to regenerate them therefrom at trivial cost. Such technology could at least arguably *falsify* the statement "All human beings should be forbidden from murder." (Simpler still: A technology that rendered human beings impervious to all physical wounds could likewise make most forms of murder impractical.)

These are some far-out examples. In other areas, though, we have already falsified many claims about state action: We have devised ways, for example, of ruling out the prohibition against practicing alternative religions. We have decided that the prohibition on the manufacture and sale of alcohol, which once seemed reasonable, is to be rejected. We no longer think that any harm comes of the attempt to perform supernatural witchcraft, and thus we do not forbid it.

In the context of American legal procedure, the provisional character of general prohibitions also manifests itself in several observable ways. Extenuating circumstances are commonly considered in criminal trials. Judges usually have some degree of latitude in sentencing. And jury nullification exists, albeit obscurely, as a remedy for exceptions that exist in justice, but that the law has failed to encode.

Yet both the presumption of liberty and the problem of falsification have additional implications for the discipline of political philosophy and its offshoot, political science.[3] When we take de Jasay's argument seriously, and when we consider the argument's Popperian implications, both disciplines should perhaps take on a more falsificationist character. A good practitioner of the science of politics would not merely set up hypotheses, one after another, that restrict the scope of human action. A good practitioner would spend a significant share of his or her time devising ways to falsify the hypotheses that justify various restrictions, and would regard existing restrictions as targets for knocking down in the future. What data might falsify them?

Falsification can be unsettling. When a claim justifying a command or a prohibition is falsified, it cannot be verified again, and the prohibition cannot remain with same rationale as before. Like falsified scientific hypotheses, falsified claims that justify state action must be rejected forever afterward. The logical form of the argument, for which much of this book serves as empirical evidence, suggests to us that the question of when to resort to the state must remain permanently unsettled, always open to reconsideration, on every point where we may formerly have concluded that use of the state is indeed necessary.

Given sufficient cultural change and technological development, might the bundle of permissible state actions one day wind up empty, not simply for strict deontologists, but for a majority, or even for everyone? Might we falsify *all* of the hypotheses that justify the finite set of commands and prohibitions that constitute the state? Is utopia an anarchy, albeit one we don't know how to achieve yet? Or is utopia an anarchy the goodness of which we do not yet sufficiently appreciate? As Lord Acton famously wrote, "Liberty is not a means to a higher end, it is itself the highest political end"—but if both he and de Jasay are right, then it is the goal of politics to make itself unnecessary.

It seems, though that we must admit that we do not know what this end state would look like: How much and which parts of ordered anarchy would work right out of the box? Which aspects would need refining over some arbitrarily long interval of practical testing? And how much of it would *never* work? Have states wrongly prevented the implementation of a full program of effective private solutions that may already exist? Or are some aspects of private, fully consensual governance still opaque to us? The literature on how an ordered anarchy supposedly would work is large and full of ideas. The experience needed to test these ideas is nearly nonexistent, which suggests that great caution is in order, yes, but also that great promise may lay ahead.

As for the future, answering the question of whether utopia has a state would seem to require insight into future technology, both scientific and social. We speak after all of solving problems *sufficiently well to rebut a weighted argument*, one that favors state action in the area in question, and not of persuading anyone to change any of their prior beliefs. When more attractive alternatives are added, one need not (and indeed should not) change any beliefs about the relative values of existing menu to change one's highest preference.

We just don't know what the new additions to our technological menu will be. For much of the history of political thought, this has not been a significant problem. Technological change was generally slow. Authors were free to invent fantastic magical devices—like Plato's Ring of Gyges or perhaps like Campanella's armillary sphere—but these remained strictly speaking incredible. Authors could also venture predictions about specific technologies that they thought might arrive at some future time, but even this type of writing has been mostly confined to the modern world, in which we have come to expect that the future will hold at least some developments of this type, if not the specific ones we have in mind. What authors have not been free to do is to make specific *and reliable* predictions about new technologies.

This inability has meaningful consequences, particularly for the more comprehensive kinds of political theorizing. Once we admit that we cannot predict future technologies, and that these technologies may nonetheless falsify our claims about the need for future state action, we are forced to admit that forecasting what utopia might look like is hopeless. The best that we can hope to do is to articulate those principles that we believe will remain true in all times and places, even despite technological change, and then argue for a set of social institutions, deploying current technologies alone, that we believe will conform best to the principles we have articulated. This effort does not allow us to say anything terribly reliable about the final end of politics, and its conclusions will always be subject to falsification.

I am, then, an *agnarchist*: I confess that for several reasons, I lack significant knowledge about how to craft the best form of society. This type of knowledge may even be permanently inaccessible, given that technology may never fully cease developing, and given that its development bears meaningfully on the question of how a society should be run. I have never been to utopia. Nor has anyone else. And nor will they be.

The idea that we might say anything meaningful about the social system of a nonexistent place, and indeed about a place that stands strangely outside the progress of technological history, is nonsense. It is further nonsense to think that anything we conclude about utopia could have serious implications for today's world. The best we can say is that we are able to articulate principles, and that some principles might be better than others, and that we are called to implement the best of them with the technologies we know how to employ today. Calling them ideal in any sense is wishful thinking.

And perhaps we are called to develop better technologies. Philosopher Jason Brennan is an advocate of laissez-faire capitalism, but he confesses that the problems of designing and implementing his favored social system are not fully solved. He writes:

> In my view, the principal problem that the capitalist ideal faces is that we do not know how to design that machinery that would make it run. The problem is ... our lack of a suitable organizational technology: our problem is a problem of *design*. It may turn out to be an *insoluble* design problem, and it is a design problem no doubt exacerbated by our selfish, predatory, and malicious propensities, but a design problem, I think, is what we've got.[4]

I agree, although this move puts him—and me—in an awkward position: We advocate a thing that we do not see as clearly as we would like. Against an opponent armed with a comprehensive social plan, one that he will confidently describe in loving and elaborate detail, our admission may even seem tactically unwise. Part of the argument of this book has been, however, that we ought to be more skeptical of such comprehensive social plans than we commonly are. Multiple powerful forces would appear to impel us in that direction, and yet these forces have little to do with the earnest search for truth.

It may prove that many or perhaps even all of the problems that we now characterize as state problems, as problems to be given to the state to solve, may in fact be *design* problems—that is, problems that could be solved in a far more morally satisfying way, if we only knew how. Proper solutions would leave behind only voluntary interactions, and possibly nothing that we might identify as a state. We just don't know how to get there yet.

Part of the answer, though, surely comes from defining our state problems as precisely as possible and from identifying commonalities among them.

For example, particularly in recent times, many state problems seem to have to do with what economists term *externalities*. Within any given regime of property rights, whether individual, collectivized, or a combination of both, it will commonly prove possible for one person's actions to affect the health, material well-being, or property interests of another, in a manner that is not priced in the market. The effect is termed a *negative externality* when the affected individual perceives it to be harmful and a *positive externality* when it is perceived to be beneficial. No system can entirely prevent these types of actions. And yet the agents have little particular incentive to refrain from negative externalities or to perform positive ones, apart from their own individual character.

In liberal democracies, many efforts to alter the legal structures defining property rights may be understood as attempts to eliminate perceived negative externalities, either by pricing them or by simply prohibiting them. The degree to which negative externalities can be alleviated through private institutions is a subject of considerable ongoing research, although it is clear that simply having a system of private property in which prices are permitted to exist in the first place is necessary for any alleviation to occur. It may be the case that finding new and more effective ways for market actors to assign prices to these events remains our most effective method of addressing them.[5]

Positive externalities, meanwhile, are to be found in all those situations that economists term *public goods problems*. Public goods are those goods that are non-excludable and non-rivalrous; this means that one can't easily prevent someone from enjoying a public good, and one person's use of a public good does not appreciably diminish the value that can be had from it by others. The term may appear to beg the question, in that it seems in all cases to call for government provision, but as we shall see, this is not so.

The trouble with public goods is simple. Because people can't easily be excluded from enjoying them, free riders emerge. Free riders benefit from the public good, but they decline to pay for it. Because public goods are non-rivalrous, adding additional people does not add much to the provider's costs of provision. As a result, individuals benefiting from a public good may find it incredible that they in particular must pay for it. After all, their use of the good isn't taking away from anyone else's. Why shouldn't someone else pay instead? Providers of the good are likely to underprovide it relative to the enjoyment that members of the public might otherwise receive from it.

Public goods problems have been known for a long time. In the *Wealth of Nations*, Adam Smith described the "obvious and simple system of natural liberty" as the one that would remain when governments avoided playing favorites in dealings with their subjects. In Smith's system, the government would have only three obligations:

> first, the duty of protecting the society from violence and invasion of other independent societies; secondly, the duty of protecting, as far as possible, every member of the society from the injustice or oppression of every other member of it, or the duty of establishing an exact administration of justice; and, thirdly, the duty of erecting and maintaining certain public works and certain public institutions which it can never be for the interest of any individual, or small number of individuals, to erect and maintain; because the profit could never repay the expence to any individual or small number of individuals, though it may frequently do much more than repay it to a great society.[6]

Since Smith's time, economists have come to realize that all three of these are public goods problems. Indeed, the last one is simply a way of restating the public goods problem itself. No one can be excluded from the benefits of national defense, and it is very difficult to exclude people from all of the benefits of domestic security, even if they personally do not make much use of the legal system. Without taxation, the state would not be able to deliver on these promises, because free riders would be more numerous than paying subscribers.

The provision of public goods is often considered to be a key reason to resort to the state. Goods such as poor relief and education are often characterized as public goods in part to solidify their theoretical claims to belonging within the state bundle. This may or may not be so, and we don't necessarily know which goods *must* be public, and which can be provided privately, but such is the claim that is made.

A famous example of this ignorance came when economist Ronald Coase went to study lighthouses. At the time, lighthouses were held to be a classic example of a public good. The argument went something like this: All ships at sea benefit from a lighthouse, whether they pay for it or not, so the good is not excludable. And when any ship benefits from a lighthouse, that ship's use of the lighthouse in no sense diminishes the benefits that will accrue to other ships. Lighthouses must be supplied by the government, because if they are not, the market will underprovide this good.

Coase, however, found that lighthouses were *anything but* public goods. On the contrary, private lighthouses had operated successfully. Through the imposition of port fees on ships as they docked, these lighthouses were able to operate as profitable concerns. The literature on various forms of private, quasi-public, and public lighthouse provision has since proliferated, showing that the problem is actually quite complex and multifaceted.[7]

In the former assumption, that lighthouses can only be public goods, we can perhaps see at work the old pro-political bias, by which it is believed that the government is or should be more active than it actually is. But we also see something more interesting: a form of ignorance that is capable of being remedied. And when it is, new possibilities open up for private, non-state solutions.

Many other examples can be found. In most jurisdictions, the state erects traffic signs and charges fines when people fail to follow them. The reasoning is, once again, that orderly traffic is a public good. While a reckless driver might get to his destination faster if everyone else obeyed the law, he would in a sense be free riding on the forbearance shown by the other drivers. We could possibly tolerate a few such drivers (indeed, we put up with them already), but if reckless driving proliferated, things would be much worse for everyone.

Meanwhile, in other jurisdictions, traffic signs have been removed from the bundle of the state, either a few at a time, or, in exceptional cases, altogether. The signs and their associated punishments have been replaced by new social technologies that curb reckless driving without as much punishment. In doing so, they offer much less opportunity for unrequited transfer. These technologies include traffic circles, calming devices like speed bumps, and the deliberate mingling of the spaces used for pedestrians and automobiles. Some towns even feature a single, altogether disarming traffic sign: "This Town Is Free of Traffic Signs," it may read.

That approach may sound dangerous, but in fact, it is much less dangerous than the status quo. When drivers know that traffic signs do not exist to guide them, and when they know that they may need to slow down for pedestrians, they take greater responsibility for their own safety. Fatalities and accidents often decline.[8] And, of course, there is considerably less use of the political means. Fewer individuals are stopped by the authorities for minor violations—which, as we have seen in recent years, can often escalate.

There is no magic to any of this. Less controlling and more effective social technologies could have been employed at any time. They simply were not, and as a result, the political means advanced. When these technologies finally were employed, the political means retreated.

Or consider finance. In recent years, computer scientists have finally developed secure digital currencies. These so-called cryptocurrencies are digital units of account that are tracked in a secure, single-instance, collectively maintained database. All users participate in the data verification process. Cryptocurrencies can do everything that state-issued fiat currency can do, and they do not suffer from commodity money's supply volatility; supply terms are typically established in advance and cannot be altered.

This in itself is no great breakthrough; gold or silver can also serve as money, as they did for thousands of years before our current fiat system. Where cryptocurrencies take an important new step is in how they are both programmable and publicly verifiable.

The latter is remarkable enough on its own as a way of preventing the sort of activity for which the state might otherwise be necessary: Proofs of payment are public and can be known by all. This eliminates a significant source of potential ambiguity, and with it will go much opportunity for inexpensive fraud through obfuscation: If you attempt a double billing, the whole world will know about it, with instant reputational effects.

Programmable currency opens up a space for entirely new ways of doing business. The release of a sum of cryptocurrency can be made conditional on consent by a third party, who can then be set up as an arbitrator. This process is nearly costless—it's just a few lines of code—and, short of cracking the cryptography, it is impossible to tamper with.[9] Payments can be issued on the meeting of certain conditions, as the arbitrator determines. The arbitrators can be known personally or they can be anonymous. They can be paid for their services, also in cryptocurrency, in the course of the contract. Arbitration can even come from a committee whose majority, or unanimity, is required for a payment to issue. Or payment might be issued if and only if the members disagree—that is, if and only if a significant dispute exists about the fulfillment of a contract's terms.

Cryptographic contracts can be built with a wide range of features, including unlinkable execution, under which the contracting parties do not know one another and will be completely unable to know with whom they contract; reputational systems, in which agents verifiably claim a public identity and are continuously associated with past transactions conducted by that identity, thereby proving that they are worthy of trust; and time released payments, whose keys only become available at some predetermined future date. Cryptocurrencies can convey digitized information as well as value, and they can be used to release this information, or not, contingent on arbitration as described above. Because of increasing network integration with the real world, even the functioning of devices can be made contingent on payment denominated in cryptocurrency; vending machines are a trivial instance of this development, and it is now no great stretch to imagine, as Nick Szabo has, that access to cars and other consumer goods might be done remotely.[10]

One possible scenario might work like this: It's the future, and driverless cars are widely available. Alice buys one using cryptocurrency. For a modest fee, she secures the services of Bob as an arbitrator for the life of the vehicle. Bob's reputation checks out as excellent; he's arbitrated many previous transactions, and he can prove it, even if, perhaps, he is unknown to Alice. Alice pays for the car and uses it happily—until one day Chuck decides to steal it. Somehow, and in defiance of the cryptographic security on the vehicle itself, Chuck succeeds, possibly by stealing a key and otherwise successfully impersonating Alice. Chuck then changes all of Alice's cryptographic keys to the car and drives it away, laughing evilly.

Enter Bob. As arbitrator, he still has a spare key, one that gives him the power to revoke all other keys—in the event of a dispute, Bob's key is programmed always to win. Recall also that Chuck could not discover this key using all the computing power in the world. Alice calls Bob, and Bob uses his key to order the car to return to Alice's driveway. He then invalidates all of Chuck's keys and generates new ones for Alice. Chuck's location is known, his keys are worthless, his reputation is ruined, and his crime is unlikely to have been worthwhile, even if no authority were ever to imprison him.

Would Bob ever be tempted to steal Alice's car for himself? Perhaps. But if he did, he would immediately and permanently forfeit his own good reputation. Alice could file a public report that the car had been taken without her consent; the fact of the transfer itself would already be a matter of record in the cryptocurrency's public database, and Bob's business would never recover.

Events like these are possible even within existing technology. They are not science fiction. With suitable modifications, Bob could even use his powers to confront the manufacturer in the event that it was charged with fraud or providing defective merchandise. (Great concern may need to be given to ensure that arbitrators do not become too close to manufacturers, particularly in industries where the latter are already few. The arbitrators' incentives must always run toward preserving their own reputational capital, and never toward one or another party whose business they are obliged to court.) Acts of theft and fraud would not stop happening in

THE FALSIFICATION OF STATE ACTION 233

the world of private, cryptocurrency-mediated dispute resolution. They would, however, become rarer, less profitable, and more easily remedied. Any potential wrongdoer would know well in advance how little his act would get him, and this all by itself would serve as a deterrent to many forms of crime. Bob's clients would be numerous, and his work would be light.

Significantly for our purposes, programmable money can do all of this without need for any direct state enforcement. The currency system enforces *all* of its own contracts, infallibly, exactly as the arbitrators decide. In short, what had been a state-provided public good, the adjudication of contracts, can now be provided wholly in private, without the troublesome question of private violence. It could even be provided, without fear, by multiple competing agencies.

If all transactions proceeded on this basis, the results would probably not be identical to those realized under a government. And they would certainly not be perfect. But to the extent that we value abstaining from the political means, cryptographic technology may do much to help. Violence, or its persuasive threat, would not be banished from the physical world, which is likely impossible in any case. But the occasions for its use in the resolution of contracts could be greatly diminished.

Some dangers would remain, to be sure. Physical threats in the real world would hardly cease to exist; but these no political system, and no private social technology, seems likely ever to banish entirely, and demanding as much may simply be unreasonable. That is, short of some fairly far-out science fiction scenarios, as mentioned above: Pervasive uploading and virtualization of the self with regular backup, for instance, might not only make some crimes more or less impossible but it might instantiate in a fully literal manner the framework for utopia proposed (though only as a thought experiment) by Robert Nozick in the third part of *Anarchy, State and Utopia*. Confronting possibilities like these would require adopting not simply a new approach to politics, but a new ethics of the self, the likes of which we have not yet conceived.

Other technologies stand in a similar relation to the state, but with the large advantage of already existing. Dominant assurance contracts are a voluntary method of providing public goods, one that is versatile and at least theoretically robust. They work like this: An entrepreneur offers a contract to a group of people who may wish to subscribe. If enough subscribers are found, each one of them must pay a previously stipulated sum. From the total, a public good is provided—a dike gets built, perhaps, or a university is founded, or some other project is undertaken that spreads its benefits far and wide. The entrepreneur takes a cut as a profit for himself. If enough subscribers are *not* found, the entrepreneur compensates those who did subscribe with a modest payoff, and the entrepreneur takes a loss.

Dominant assurance contracts look a lot like the now-common practice of *crowdfunding*, which has rapidly grown into a significant source of funds for art and entertainment.[11] The difference between crowdfunding and a dominant assurance contract is that in the latter, subscribers *always* stand to gain something. If the contract succeeds, they gain the provision of a public good, at the price that they agreed to pay. (Note that the "good" of a public good almost always falls unequally on various people; presumably each subscriber deems the amount that they will personally capture to be worth the subscription price—and thus subscribing is a good deal for them. Free riders are welcome to come along too, if it so happens.) Or, if the contract fails, subscribers could just get a monetary reward on what turned out to have been a bond-like investment rather than a purchase. But either way, they've won something, which should make the contracts appealing to them in principle.

Game theoretical analysis by the economist Alexander Tabarrok suggests that these contracts should work in a wide variety of circumstances. Further, a contract's failure will reveal important information about public preferences; a failure will allow us to make inferences about the actual value of various public goods, as assessed by the public, and this may lead to more appealing contracts in the future. Until recently we have had few effective means of determining these preferences.[12]

In none of this has the government played much of a role, except for its basic function as an adjudicator of contracts—and again, that's a role that might be played by a private arbitrator. Dominant assurance contracts have even been implemented in the Bitcoin system, which would render the adjudication process completely private. Public goods may not need to be public after all.

What's perhaps most interesting about the dominant assurance contract is that it is a social technology that might have been invented a long time ago, but wasn't. The only necessary concepts are standard contract law and a rudimentary notion of public goods. As we've seen above, such knowledge was already common in Adam Smith's day. But dominant assurance contracts were only invented in 1998. They thus raise an intriguing question: What *other* social technologies are out there, waiting to be discovered, that could make the world a more voluntary place?

Like cryptocurrencies, dominant assurance contracts seem poised to falsify significant swaths of our existing justification for government. Might technology eventually render all government provision of public goods superfluous? Impractical? Even *impossible*? As I have argued above, I do not believe that we can answer these questions. I do think it behooves us to find alternate solutions to the problems for which we now deploy the state. Presuming that the state is the best provider, and that it must always remain the best provider, seems like quite a leap in light of how little we know. I suspect that dominant assurance contracts, programmable currencies, and other emerging social technologies should inform political theory in a way that they have not yet done so far. It may be deeply annoying to consider that perhaps *no* political theory has ever been sufficiently informed, or sufficiently armed with alternate social technologies, to craft a good society. But one should never reject a prospect simply because it annoys.

The point of illustrating several of these solutions is not to suggest that they alone would suffice, and that the bundle of state problems can be emptied today in ways that you or I would always find acceptable. My purpose here is twofold. First, it is to conceptually unite some very interesting developments in the social sciences and in information technology. The connection they share is that they promise to remove to some degree the need to resort to violence, or threats thereof, to attain important public goods. Second, it is to cast these developments as part of an ongoing field of inquiry, one that deserves much more attention than it has received so far, even from libertarians and classical liberals. This field takes as its object the attempt to provide substitutes for government action, with the understanding that these substitutes can solve more than merely practical problems. They can also alleviate for us the ethical dilemma of government itself. Better solutions are available, perhaps, than we have so far been aware of. It is incumbent upon us to find them.[13]

In the meantime, admitting that we do not know what utopia will look like will save much effort that would otherwise be wasted (and has been wasted) on a series of sterile intellectual constructions and equally sterile debates about them. The concession also regrounds us in the search for present-day public policy solutions. Libertarians and classical liberals in particular have a distinct role to play in democratic and economically mixed polities, one that is unlikely to be played by the adherents any other ideology: It would appear that we alone are likely to suggest wholly new ways of solving problems with less reliance on the state, and to point out

when and how non-state or less-state solutions may do better. If we can
do that, then perhaps the others will listen.

NOTES

1. In the paragraphs that follow, I adhere closely to de Jasay,
   "Freedom, from a Mainly Logical Perspective," in *Political
   Philosophy, Clearly*, pp. 206–227.
2. Popper, *The Logic of Scientific Discovery*, p. 18.
3. As opposed to political philosophers, political scientists may protest, with some justification, that they *do* engage in falsificatory
   work. And yet it remains frustrating that, in the context of political
   argument, "this probably isn't going to work" seems to have less
   evocative power than "I have found the way to salvation." In this,
   the philosophers seemingly hold the whip hand.
4. Brennan, *Why Not Capitalism?*, p. 40. The passage parodies a
   nearly identical one from G.A. Cohen, *Why Not Socialism?*,
   pp. 57–58. Design problems would appear endemic to many highly
   theoretical political systems.
5. One particularly interesting recent book in this area is Stringham's
   Private Governance.
6. Smith, *The Wealth of Nations*, 1776. Book IV ch IX, http://www.
   econlib.org/library/Smith/smWN19.html.
7. Coase, "The Lighthouse in Economics." See also Barnett and
   Block, "Coase and Van Zandt on Lighthouses," which includes an
   overview of the literature.
8. Vanderbilt, "The Traffic Guru." *Wilson Quarterly*, Summer 2008.
   http://www.wilsonquarterly.com/article.cfm?AID=1234.
9. "Cracking a 4,096 bit RSA key with best known algorithm really
   would require more electrical power for the computers than the
   power produced by a supernova." To say nothing of securing the
   processor time. Szabo, "Formalizing and Securing Relationships
   on Public Networks," available at http://szabo.best.vwh.net/for-
   malize.html.
10. Ibid.
11. Boyle, "Yes, Kickstarter raises more money for artists than the
    NEA. Here's why that's not really surprising." *Washington Post*,
    July 7, 2013. http://www.washingtonpost.com/blogs/wonk-

blog/wp/2013/07/07/yes-kickstarter-raises-more-money-for-artists-than-the-nea-heres-why-thats-not-really-surprising/.

12. Tabarrok, "The private provision of public goods via dominant assurance contracts."

13. Stringham, *Private Governance,* argues that private governance has played a key role in crafting the rules that govern a wide array of social and economic activities. The state is a last resort, much as has been argued here. Stringham likewise resists what he terms "legal centralism," that is, "the idea that order in the world depends on and is attributable to government law" (p. 5).

CHAPTER 12

Advancing Technology Demands Intellectual Modesty

It would be unreasonable to consider technology's liberating potential if we did not also consider, at least briefly, its capacity to enslave and oppress. Let us do so now. Beginning with the advent of modern statistical record-keeping, and accelerating with the information age, states have increasingly had the means to enact widespread population-level surveillance. This surveillance is a prerequisite for many authoritarian social interventions. Even Plato seems to have been aware of the need for something like it, what with his insistence on a never-changing social structure, composed of a fixed number of families, with a fixed number of men always holding various offices and occupations. *Someone*, presumably, would have to watch carefully and make sure they all stayed there.

As long as effective technologies of population surveillance did not actually exist, the theorists of authoritarianism could have their cake and eat it too. They had free play to elaborate a wish list of things that they might implement, if only they had the means. But on the margin, this wish list was relatively free of consequence. Adding another item to the list of things that the government would (always competently) manage in utopia would make utopia's government look better, but it would never make any actual regime look worse. From the ancient world to the industrial era, thinkers freely invented total systems and devised radical projects for the remaking of mankind, largely without the risk that these projects might ever be put to the test. They all demanded more in the way of social control than could be brought to bear in the real world—but at least for them, it didn't matter.

© The Author(s) 2017
J. Kuznicki, *Technology and the End of Authority*,
DOI 10.1007/978-3-319-48692-5_12

One wonders, for instance, exactly how the Solarians might have been constrained to mating only every third night, and only with the partners appointed to them, and how all other expressions of sexuality would have been suppressed. And this is just one example from a near infinity of others that might be offered about social planning in Solaria. The system can only appear to work when we do not consider too closely the mechanisms that would be required to ensure obedience. Whether these be radically reshaped personal values that leave absolutely no room for sexuality as an expression of love, or a pervasive electronic surveillance over sexuality, or just a vigorous secret police, no mechanism toward this end is terribly appealing or plausible as part of a purportedly *good* social system.

Perhaps one of the reasons why Campanella's utopia seemed worth thinking about—and Plato's, and so many others—was precisely that these utopias were unattainable, and even mostly unimaginable, in practice. They were beautiful dreams, but they were only beautiful as long as they remained obscure. The technologies that would make them possible would simultaneously make them ugly; but one never needs to consider these ugly aspects when one constructs a city only in words, as Plato and Campanella both did.

Making people comply in practice with highly artificial commands requires technologies of pervasive social control that only began to arrive in the industrial era—things like the panopticon, automated recordkeeping, audio and video surveillance, biometric identification, and the like. We may now try for a meticulously planned utopia in ways that no one previously could. In the modern era, the philosopher kings have in a sense come into their own.

The results have been horrifying. Technological developments have enormously empowered social engineers, and yet it has always been at the expense of their fellow citizens. Again and again, the twentieth century built a Bakunin's Throne and placed someone on it—and tens of millions died before it. The industrial power that commands the transfer of vast supplies of food into cities is also the industrial power that entirely deprives a countryside of food. The story of industrialization's role in the Holocaust is likewise altogether well known, and with good reason.

The story of political theory's technological interface with practical politics may even be summed up as follows: For centuries, we told ourselves stories of comprehensive social control, until finally it appeared that we had the power to make them real, which we promptly did. And after only a few years, we learned to stop telling those stories, because they led nowhere good. The emergence of a countervailing cultural tendency—the rise of consciously dystopian literature—fills me with a paradoxical hope. Dystopias were rare prior to World War II, but in any survey of today's fiction, one can hardly escape the idea that comprehensive social planning in the service of any end whatsoever is a short path to disaster. *Brave New World*, discussed above in the course of examining utilitarianism, is but one example of this tendency, although it's an early and important one. *1984* of course is another, and, with *The Hunger Games* at their head, there are now so many dystopias written for young adults that young-adult dystopian literature constitutes something of a genre unto itself. We may be tempted to laugh at or dismiss the vulgarization of any intellectual trend, or even just its commercialization. We know of no other reliable means, however, by which the ideas of educated people propagate from their origins.

I can only hope that we are becoming culturally inoculated against our former utopian tendencies. What is curious about all of these dystopias, and what is important in their emergence for the history of political thought, is that they are essentially visions not of what is right, but of what is (or could be) wrong. As a necessary consequence, dystopian literature is more *intellectually modest* than its utopian counterpart. Articulating a type of failure does not require anyone to articulate exactly what success would look like, and most dystopias do not. Dystopias are thus apt to hit the mark in ways that utopias (or just positive political programs) no longer seem able to do: few find George Orwell's anarcho-syndicalism as compelling as his searing, all-too-real portraits of totalitarianism. Aldous Huxley's final, utopian novel, *Island*, is generally forgotten, though no one will forget *Brave New World*.

In the realm of theory as well, significant factions of western political thought have lately taken a giant step back from the elaborate governing projects that have animated so much of the discipline. From the Frankfurt School on the socialist left, to the libertarian Austrian School, to Oakeshottian conservatism on the right, political thought has retreated from centralization and from the promise of comprehensive control. The presumption has shifted decisively away from the idea that the good society is simply a matter of proper planning at the center. The center itself may be the problem.

Even when we consider the most prominent in the field, it seems clear that political philosophy has backed away from the building of cities in words, and that it has lately attempted simply to define some of the abstract preconditions that might be necessary, albeit not sufficient, for creating an ideal society. Robert Nozick's framework for utopia rather obviously comes to mind. It is described as follows:

> Imagine a possible world in which to live; this world need not contain everyone else now alive, and it may contain beings who have never actually lived. Every rational creature in this world ... will have the same rights of imagining a possible world for himself to live in (in which all other rational inhabitants have the same imagining rights, and so on) as you have. The other inhabitants of the world you have imagined may choose to stay in the world which has been created for them (they have been created for) or they may choose to leave it and inhabit a world of their own imagining .... This process goes on; worlds are created, people leave them, create new worlds, and so on.
>
> Will the process go on indefinitely? Are all such worlds ephemeral or are there some stable worlds in which all of the original population will choose to remain?[1]

Nozick asserted that *stable* worlds—that is, those which no one ever chose to leave—would also be worlds in which no one could even *imagine* a world that they thought would be better. In keeping with one of the themes developed in this book, I might ask whether changing social and technological conditions inside any stable world might eventually lead to its instability, in the narrow sense that Nozick uses the term, and if so, what that might say about the notion of stability itself. But one thing is clear: The framework for utopia offers little specific guidance for social institutions; it is, rather, a proposed means of evaluating such institutions. It is not Plato's *Republic*. Instead, it is a question about Plato's *Republic*, to wit: Who, exactly, would want to live there? And under what conditions of entry and exit?

And of course there is John Rawls, who is commonly regarded as the pre-eminent political philosopher of the twentieth century. It is difficult to write about political philosophy without paying at least some attention to his work. Admittedly, however, I approach Rawls with reluctance: The literature is vast. Almost everything in the field today is seemingly a commentary on Rawls, a tendency I have sought to avoid. Many of Rawls' own students are still living, even, and the case may be made that Rawls doesn't belong to historians like me just yet.

Still, I believe two points can be made about Rawls' work. The first is that it tends in part to support the thesis I have just advanced, namely that political theory has begun to stage something of a retreat from the presumption of the state's centrality and necessity. This may seem a strange assertion, but I will happily and briefly defend it.

The second point that should be made about Rawls' work is that it still greatly overrates *stability* as a trait of good societies. Rawls' work overvalues stability, it would appear, in part because it gives scant consideration to the prospect of technological change and/or the likelihood that technological change can meaningfully bear on the question of which social institutions are preferable. The overrating of stability is also likely an artifact of a philosophical consensus that stretches back to Plato, which tends to conflate the good with the stable, and onward from there to the eternal. As we have already had occasion to note, however, there is no especially clear reason why eternity is a quality we should prefer, ceteris paribus, in a state, or indeed in any set of social institutions.

Let us begin with the relatively modest place of the state, that is, the modern locus of government, within Rawls' work. A careful reading shows decisively that Rawls did not intend to found a polity in words, as so many of his predecessors did, and much less did he intend to found a modern state in words. In section 3 of *A Theory of Justice*, Rawls writes:

> My aim is to present a conception of justice which generalizes and carries to a higher level of abstraction the familiar theory of the social contract as found, say, in Locke, Rousseau, and Kant. In order to do this we are not to think of the original contract as one to enter a particular society or to set up a particular form of government. Rather, the guiding idea is that the principles of justice for the basic structure of society are the object of the original agreement.[2]

This original agreement is to take place without reference to personal circumstances of any kind whatsoever; principles alone are to be considered. Famously, all else is hidden behind a veil of ignorance, at least for the time being. Rawls adds: "This original position is not, of course, thought of as an actual historical state of affairs, much less as a primitive condition of culture."[3] It is a framework for thinking, not an epoch. This move places him closer to Locke and Kant than to Rousseau; as we have seen, Rousseau closely identified the state of nature with a particular—albeit imaginary—era of human development. Unlike Locke, however, Rawls is quite explicit that reasoning conducted in the state of nature should not aim directly to set up a government. Rather, such reasoning aims to set forth abstract "first principles of a conception of justice" that are to be binding on all. Only later will the first principles be used to evaluate various proposed social institutions, forms of government included.[4] For Rawls, the state of nature serves to banish "the accidents of natural endowment and the contingencies of social circumstance" from our thinking about justice. One such accident may even be the sheer fact that we take the existence of government for granted. At least this much is certainly implicit in the abstract approach that Rawls takes to the state of nature.[5]

Following an examination of various proposed principles of justice that we need not recapitulate here, Rawls settles on two of them. Briefly stated, they are, first, "equality in the assignment of basic rights and duties," and, second, that "social and economic inequalities, for example inequalities of wealth and authority, are just only if they result in compensating benefits for everyone, and in particular for the least advantaged members of society," and he is careful to rule out the utilitarian demand that institutions should be justified "on the grounds that the hardships of some are offset by a greater good in the aggregate."[6]

It must be said that neither principle immediately compels the formation of a state. Indeed, acts of governing are necessarily subject to Rawls' second principle, for such acts all produce inequalities of authority. As such, they must be shown to work for the benefit of the least advantaged. If not, they must be rejected. Much has been made of how Rawls' second principle may serve to justify a welfare state. Not enough, however, has been made of how Rawls' second principle otherwise indicts the expansive status and authority of the state, at least in many other matters: Offhand, it clearly challenges much about the needless pomp and inflated social rank of those who conduct the modern state; it challenges contemporary police procedures, which seem clearly rigged against the disadvantaged; it challenges American mass imprisonment in particular, notably the plea bargaining system and its unfair use against the indigent; it further challenges the criminalization of narcotics, the seizure through asset forfeiture of the holdings of ordinary individuals, and the criminalization of small-time street-level entrepreneurship, exactly as it is often conducted by the least well-off. To a proper Rawlsean, the government of the United States among many others should have a lot of explaining to do. That Rawls' central argument has generally been used to prop up the state may one day be regarded as surprising.

The road to particular institutions is long, or at least it should be. In an elaboration of his argument, Rawls writes that parties in the initial position will first agree on the principles of justice while divested of all knowledge of particular circumstances, including the workings of political institutions. "Thus I suppose," he writes, "that after the parties have adopted the principles of justice ... they move to a constitutional convention. Here they are to decide upon the justice of political forms and choose a constitution ... the veil of ignorance is partially lifted ... they now know the relevant general facts about their society, that is, its natural circumstances and resources, its level of economic advance and political culture, and so on."[7]

Theorists ever since have argued about exactly which facts should and should not count at this stage of consideration, and the debate is unlikely to conclude anytime soon.[8] One thing that must be observed, however, is that this sort of debate, this debate about the proper role of natural circumstances, resources, economic development, and political culture is partly an *empirical* debate. It's a debate, then, that may contain areas in which all parties are sincere and impartial, and yet still in error with regard to the facts, whether through believing generally accepted falsehoods or through sheer ignorance. At this stage of the debate, we likely all "know" things that are plain wrong, and this will bear significantly on our choice of institutions. We also likely don't know, and can't even imagine, certain truths that would be relevant to our choice, were we to discover them.

Rawls acknowledges this difficulty at several stages in his argument, admitting as noted earlier that to avoid evaluating interminable lists of possible social principles and institutions, he will confine his investigations to "a short list of traditional and familiar principles."[9] A searching evaluation of such a limited list may leave few members standing, but it is unclear that the winners in this first round of judging will continue to fare so well indefinitely. Conceivably, future societies may contain institutions the likes of which we have not dreamt of, and these may be superior, on Rawls' terms, to our own. Also conceivably, these future societies may only arrive at their new institutional arrangements through technologies that we do not yet possess. Even if the abstract principles of justice are eternal, the conclusions at the second stage of Rawls' reasoning, the stage that is likened to a constitutional convention, may be mutable.

Rawls is loath to admit such mutability; he writes that at the level of principles of justice, "None ... is contingent upon certain social or other conditions. Now one reason for this is to keep things simple. It would be easy to formulate a family of conceptions each designed to apply only if special circumstances obtain, these various conditions being exhaustive and mutually exclusive."[10] Arguably Marx's stadial theory of history did exactly that, and others perhaps have done likewise. Rawls deems this approach "very complicated if not unmanageable," however, and he abandons it, positing that for any such stadial theory, some underlying principle must surely supply its reason for being, and it is this that we must examine instead.

And yet not only Marx but many others have proposed stadial theories of history, and it is odd to see all of them dismissed as too complicated to examine. Others have certainly examined them. Against this widely shared view, Rawls insists that a well-ordered society—one that significantly resembles a fully just one, if not one that fully instantiates justice—will tend to be stable:

> [A] well-ordered society endures over time ... when institutions are just ... those taking part in these arrangements acquire the corresponding sense of justice and desire to do their part in maintaining them. One conception of justice is more stable than another if the sense of justice that it tends to generate is stronger and more likely to override disruptive inclinations ...[11]

Apart from the popularity of stadial social theories, two objections seem obvious: First, those taking part in manifestly unjust arrangements seem also to have a desire to maintain them. They too wish to see their institutions perpetuated, particularly if they are the beneficiaries, or if they fear punishment from some new regime. These considerations should not be raised from behind the veil of ignorance, of course, but it can be difficult in practice to distinguish when stability is rightfully urged and when it is not: even if justice inculcates stability, stability—in the sense of preferring the maintenance of existing institutions—by no means tracks justice at all. And second, we have no reason to believe that the social arrangements identified by Rawls are stable *because of their justice.* These arrangements, recall, are drawn from a traditional list; they may be the winners from among that list, but we have no guarantee that they are the winners in an absolute sense, and thus any stability they may have does not necessarily come from their justice. It may come from their injustice, or from some neutral factor, like fear of the unknown, or ignorance of the alternatives. Below I will present some evidence that our list of alternatives is much smaller than it should be, and that the design space we consider is much larger than traditional social theory, Rawls included, has allowed.

Ultimately what Rawls has supplied us is a set of decision-making criteria, combined with its application to a partial list of alternatives.

Both the criteria and the alternatives may be incomplete, however, and I believe that to a significant degree they are. And yet I must confess that there is something much preferable to Rawls' incompleteness when we compare it, say, to Plato's strutting but obviously misplaced certainties. When the charge is raised that philosophy never makes any progress of the sort that is seen in the natural sciences, one might do well to compare these two philosophers, and to weigh the value of a hard-won humility.

But what remains of political theory? What of the canonical texts and their grand ambitions? Isn't it a rather depressing prospect to consider how little the discipline may really have to offer? More and more, a theorist's choices appear to be as follows: First, recommend totalizing legislation, which can and will be enforced oppressively, owing to industrial and information technology. Second, recommend specifically *partial* legislation, which will be enforced sporadically, create ambiguity, allow some disobedience, and yield prejudicial enforcement, which is hardly defensible in ethical terms. And third, recommend *no legislation at all*, repeal all or at least the worst of what stands and awaits whatever consequences will follow. In any particular topic that political theory may address, these three may be our only choices. One may fairly object that none of them are especially inspiring.

The first two choices are especially unpalatable. It is probably not the case that advanced technology will ever deliver fully effective compliance with totalizing legislation. All life is an arms race, and technologies have at times advanced both liberty and oppression. Whenever technologies arise to control, others will arise to subvert control. An equilibrium emerges between them, sometimes inclining more to one side, and sometimes more to the other. This has been a constant, I believe, throughout history. I infer, I hope reasonably, that this state of affairs is unlikely to change. But within this framework, technology could still stand to make government much more unpleasant in the future, even if the guiding vision of a totalitarian society is never realized. Of course, *this too* is a prediction about future technological advances, and, like the hopeful predictions of technological liberation, it cannot be made with great confidence.

Ambiguity and partiality in the law have always been unacceptable to everyone, at least when stated formally. It is always possible to advocate them—what is good for General Motors is good for America, say, or whatever pleases the king is law—but such moves are rightly derided as incompatible with the philosophical way of thinking. The final alternative, to recommend no control at all, now suggests itself with more and more urgency, whether it is inspiring or not.

Recommending no legislation at all, and repealing what we know how to repeal does not mean that the political theorist's work is necessarily done. This path is not a unity; there are still choices to be made among countless alternatives, and evaluating these choices too might be considered a part of the political theorist's calling, albeit a neglected one. Should it not be strange that, for a whole branch of philosophy, an idea is only properly on the table if it's not voluntary? "The design and administration of non-voluntary mechanisms of collective action" may be a valid description of politics, but if so, shouldn't the adjective "non-voluntary" be its most problematic aspect?

As discussed above, the state is a bundle, and expansive claims about state power are open to doubt on many fronts. The state may not even have any well-defined ends for which we can know with certainty that it is the best possible means. We usually have at least some reason to suspect that it is not.

But for exactly the same reasons, our attempts to craft political programs are *also* suspect. Just like the state itself, our ideas about what the state *ought* to be doing are also bundles made up of uncertainties. Political programs are cobbled together of makeshift parts. All of them are subject to structural forces, and all of them suffer from similar knowledge problems.

It is easy to understand why this should be. At their best, political programs are formed by inducting across incredibly vast fields of evidence, and the evidence is apt to betray the activist when he least suspects it—as when Engels noted, with some confusion, that the living standards of the English working class had actually been rising during his lifetime. The project of declaring what we should do politically is always an inductive endeavor concerning a complex world, about which we know all too little. It attempts, with what one might call great immodesty, to gather up many different particular facts and then draw general conclusions from them. Even when it is articulated in idealized and deductive terms, as in much of rights theory, the art of application makes inductive demands on the practitioner whenever a particular case is to be examined.

Faced with the flawed human world, political theories and political programs may propose to avoid the appearance of being based upon it at all. Many have claimed to rest on divine revelation instead, or on self-evident truths. But even if so, we are entitled to ask how we can expect any good results from them, given that they begin such a great intellectual distance from the thing that they propose to order. How are we to apply self-evident or divine truths to the world, if not by a process of induction that is every bit as fraught as the one that would be required for a system whose claims were grounded on empirical observation? Where exactly are the angels who will administer and comply with whatever the theorist has proposed?[12]

For empirically based theories, meanwhile, the problem remains that history happens only once, in one direction, and there can be no controlled experiments upon it. In this, political theory is quite different from physics or medicine, in which repeated trials are more easily had. And yet *history* remains the data on which political theory must perform its inductions.

As a result, the end product of political theorizing ought to be a highly constrained set of abstract principles, mostly cautionary in character, and one that we have comparatively less reason to trust overall. There is nothing necessarily wrong with the attempt at crafting such abstractions; in many cases, we must theorize, and generalize, simply for reasons of mental economy. Taking in the whole picture will always be impossible, and yet acting on the whole picture will in some sense always be obligatory; even declining to act reveals a preference, as it were.

And abstraction can be powerful. Consider Frederick Douglass' ideology of self-ownership, and the arguments it produced against slavery and the subjection of women. Or John Rawls' original position, which is both an abstraction and a useful tool for thinking about many different social arrangements, both governmental and otherwise. Or John Stuart Mill's harm principle. Each is admirable, even if each faces clear limitations. All of them may be tools in a thinker's armamentarium. But none may be the *only* tool.

Programmatic ideologies in particular usually age badly. Consider the following policies: the graduated income tax, immigration restrictions, eugenics, women's suffrage, and the prohibition of alcohol. All were strongly associated with the American progressive ideology in the first decades of the last century. In their collective heyday, these ideas seemed to have a unity about them, and partisans could often be found defending all or nearly all of them together.

For good or ill: Today almost all Americans find the graduated income tax acceptable, immigration restrictions right-wing (and not left-), eugenics unconscionable, women's suffrage morally obligatory, and Prohibition a terrible mistake. Lashed together for a time, these positions have had anything but a common fate ever since.

The point here is not to condemn, or to praise, early twentieth-century progressives. It is to note that when we look at readily identifiable, real-world political belief sets from the outside, they often turn out to be products of the times and places in which they were crafted. They were brought together by one political moment, only to be separated by the next.

This is very different from how they look to a true believer. To those who hold a belief set, its contents may appear entirely correct and coherent. As a whole the belief set may even appear to spring from a deeper, even a transcendent source. Or at least, so says almost everyone who holds strong political beliefs. Almost everyone is probably wrong about this.[13] What feels like a unity in our political beliefs is probably an illusion, and it will all but certainly be regarded as such by the future.

It may be satisfying, but I think too easy, to say that relativism itself constitutes the one true ideology, or at least the one true response to all others' ideologies. The difficulty here is that real-world problems don't go away when we make that move, and relativists must still come up with solutions to them. And these solutions are still based on giant inductive leaps from huge and barely understood real-world phenomena. Sweep away all systems, and the result is a system. One that is likely held with even more self-satisfaction, and maybe even less self-criticism.

We may even know all about the play of contingency—and *still* find a given ideology compelling. This is roughly the position that I occupy. It has not escaped my notice that in all probability, two hundred years' time will leave many of my cherished beliefs in tatters. A few might become foundational. The remainder will be thrown to the winds. I do hope none of them are found unconscionable, but again, there's not much to be done about that. It consoles me only somewhat that the same fate will befall everyone else's cobbled ideologies, and that those who will suffer the most in history's judgment will be precisely those who have not bothered to acknowledge their difficulties here. To hold an ideology responsibly is to be aware of how it stands to be unmade in ways we cannot know yet.

Thus while we may all be sincere, we should all learn to wear our ideologies lightly. Not to discard them, not to declare them untrue and not to cynically abandon all belief in the quest for a post-belief nirvana. But rather: We must say that these are the things we believe, regardless of what history will do with them. And though time may prove them wrong, it has not done so yet. Here I stand, *until you convince me otherwise.*

The application of political theories to the world should be far more modest than the applications of scientific theories, insofar as the former are not so easily based on controlled experiment. Yet as we have seen, the reverse is often true. Political theories are commonly pursued with a fervor and a single-mindedness that greatly outstrips the theories of science. We hold our political beliefs in a way that might shock physical scientists, and perhaps leave them a bit embarrassed, if they imagined such expansiveness invading their own domain. Do genuine scientists of any variety ever hold pep rallies for their views? More to the point: How often do they kill for them?

Let us resolve to have a politics shot through with doubt, so that, if it ever comes time to do murder for our politics, our very opinions about politics will make us hesitate, long and hard, before pulling the trigger. Let us be *meta-rational* about our politics, and recognize that this is an area where we humans have constantly gotten things wrong, and where we have constantly killed and died in vain. Let us adopt a world-view that accords well with our well-known human failings. Let us tell ourselves—hopefully with all the allure of an ironclad certitude—that we are prone to being wrong, and that it is ghastly to kill for a mistake.

With others in the same position, we can share a bond of camaraderie, regardless of our particular conclusions. We can know the future may make fools of us all, and that in all probability it will. But we don't know just how the future will do it, and anyway, we are still permitted to believe, as long as we do it modestly. If our beliefs are to be overthrown by something better, that happy event will arrive only because you and I have earnestly fought the good fight in the present. And it's a fight in which, at any rate, we can now join sincerely and without fear.

NOTES

1. Nozick, *Anarchy, State, and Utopia*, p. 299.
2. Rawls, *A Theory of Justice*, p. 10.
3. Ibid., p. 11.
4. Ibid., pp. 11–12.
5. Ibid., p. 14.
6. Ibid., p. 13.
7. Ibid., pp. 172–3.
8. See, for example, Lomasky, "Libertarianism at Twin Harvard."
9. Rawls, *A Theory of Justice*, pp. 106–7, 294.

10. Ibid., p. 108.
11. Ibid., p. 398.
12. See, for example, Levy, "There's No Such Thing as Ideal Theory" (April 3, 2014). Available at SSRN: http://ssrn.com/abstract=2420125 or http://dx.doi.org/10.2139/ssrn.2420125.
13. The most important exception I can think of is Robert Nozick, who often confessed to varying degrees of certainty about different parts of his intellectual project. He also wondered why others didn't do the same. Nozick was probably more right about this observation than he was about anything else he ever wrote.

CHAPTER 13

On Trade as a Central Feature of Society

If, as I have come to believe, the state is not central to society, is something else central? It is unclear that anything qualifies. The center is a metaphor, after all. What's worse, it's a metaphor that seems to impair our thinking. We are never obliged to employ any particular metaphors, and perhaps we are best off without this one.

Yet many aspects of society are, if not central, at least not easily done away with. They would appear to be intrinsic features of human life, regardless of what governments claim to think or do about them. One such feature may be *trade*.

It has sometimes been asserted that communism and capitalism are asymmetrically situated with respect to one another: Under most forms of communism, a capitalist business enterprise will face severe constraints. Such an enterprise probably cannot operate lawfully, of course, and some types of enterprise may not be able to function at all. Business operators stand to be brutally repressed, and their families might suffer likewise.

Meanwhile, under a capitalism, it is typically quite easy to set up at least some forms of communal property. Although these are not identical to the full, society-wide instantiation of communism, one may fairly object that a single small-scale capitalist business enterprise stands analogously; it, too, is not identical to the full, society-wide instantiation of the social system that it might otherwise call home. But under a capitalist social order, as long as the proceedings remain consensual, the state will generally honor participants' desires and recognize voluntary property transfers to the collective in the ways that they stipulate. Little would appear to prohibit a

© The Author(s) 2017
J. Kuznicki, *Technology and the End of Authority*,
DOI 10.1007/978-3-319-48692-5_13

communism from emerging spontaneously and peacefully under capitalism, while the reverse transformation could not occur, at least in most communist societies so far witnessed, so long as the state system of repression remained intact.

Thus if a group of dedicated communists somehow found themselves living in a fully capitalist society—whatever *that* may look like—they could presumably pursue their communist ideals all the same. In Leonard Read's famous phrase, they would be allowed to do "anything that's peaceful." A capitalist society could tolerate communist enclaves, but not vice versa.

It is a worthwhile asymmetry to consider. Let us return though to the capitalists who wish to behave capitalistically under communism, in an enclave with no restrictions on trade or on the owning of private property. Setting aside the obvious danger of repression, let us consider the incentives that would operate.

First, large-scale markets depend on a widespread respect for private property rights. Without this respect, markets of the scale we are used to will never emerge. Markets of any size will also need reliable mechanisms for settling disputes, or for making sure that disputes do not arise in the first place. Ideally, these mechanisms will contribute as well to a respect for the overall system of property. A market-respecting culture need not be perfect in judgment or unfailing in execution to realize substantial gains from trade. But without a reasonable expectation of the enjoyment of the fruits of property, the market process will not go forward very efficiently.

Our market under communism would therefore necessarily face severe limits, even after we set aside the threat of persecution. The market's activities would have to remain small in scale, and probably limited to a circle of trusted participants. Dispute resolution would be a constant problem. Disputes do tend to go public, after all, whether one wants them to or not. This market would therefore need to keep disputes of all kinds to a minimum. It would probably be limited to trading only a small roster of commodities whose quality and quantity could be easily verified. Many common aspects of the market as we know it, including specialized labor, advertising, public auction, banking and finance, and trade in securities, would be extremely difficult if not impossible. The entire process would be considerably less efficient in its realization of gains from trade, relative to markets in societies where they enjoy greater institutional support.

Under such conditions, we might nonetheless have a market of sorts. And, as it turns out, people living under communism have *always* conducted small-scale markets of exactly this type. Trade doesn't actually disappear under communism. It simply shrinks in scope and realizes much smaller rewards. But it remains indispensable, and omnipresent, at least if practical evidence is to be trusted. As James C. Scott observed in his stimulating book *Two Cheers for Anarchism*:

> In one typical East German factory, the two most indispensable employees were not even part of the official organizational chart. One was a "jack-of-all trades" adept at devising short-term ... solutions to keep machines running, to correct production flaws, and to make substitute spare parts. The second ... used factory funds to purchase and store desirable nonperishable goods (e.g., soap powder, quality paper, good wine, yarn, medicines, fashionable clothes) when they were available. Then, when the factory absolutely needed a machine, spare parts, or raw material not available through the plan to meet its quotas and earn its bonuses, this employee packed the hoarded goods in a Trabant and went seeking to barter them for the necessary factory supplies. Were it not for these informal arrangements, formal production would have ceased.[1]

Trade is so common to human beings, and its advantages at the small scale are so obvious, that it would appear to continue in a limited form even under intense oppression. The very success of communism, such as it was, apparently depended on the fact that the repression of trade was so often ineffective. Practically, it could not be otherwise, because it would appear often to have been the only thing that made the factories run at all. It has recently been argued that small-scale trade has become quietly omnipresent even in North Korea, which may have the most repressive state now in existence.[2]

This, then, is another sense in which communism and capitalism are asymmetrically situated: Trade exists in both, and both depend on it. But only communism attempts to destroy the thing that its material success depends upon. Trade is an action that proceeds *as if there were* private property. Transferable property claims that endure over some interval of time are *implicit* in the act of trade, and one cannot even say what trade is without invoking them.

It is also well established that ownership-like behaviors, including even exchanges, occur in the animal kingdom. Animals behave as if they own territories, nests, burrows, and food stores. There is thus little reason to expect that the disappearance of trade and property will come easily, regardless of what our political theories command us to think about them.[3]

Given that trade will probably never go away, the question now becomes how various social systems react to its presence. Do they encourage trade or suppress it? Do they channel it, and if so, to what ends? Which goods and services will be considered legitimate objects in trade, and which will not be? What rules will govern trade, and what incentives will these rules create? How are ownership claims asserted? How are they maintained? And how are disputes settled?

All are weighty questions, and we cannot discuss them with the depth they deserve. But when your neighbor has only gin and you have only tonic, certain ideas inevitably start to form, and it would be a rare communist indeed to refuse such tempting notions on the grounds that those who barter behave as if they own private property. Pragmatically, collective ownership under modern communism has usually not extended to consumption goods. (Although for Plato and many others, it would have.) As we have seen from the example of the East German factory, the limits of collectivization in practice would appear to be set by the functional limits of state repression, and not by any particular ideological claims: People are only as communist as the state can compel them to be.

We may do better, in thinking about communism, to consider that "communism" is a veneer rather shoddily laid over a deep and natural substratum of human behavior that necessarily entails the assertion of private property claims. This substratum never really goes away, it would seem; it is only covered up or redirected for a time, and governments may be judged good or bad insofar as they do better or worse at reckoning with an ineradicable desire, the desire for trade in pursuit of mutual advantage.

A market-oriented economy in practice will view such trades as a first resort. It will typically identify in law various types of property, the means by which they are secured, documented, and conveyed, and how various property rights disputes are to be adjudicated. A market-oriented society will acknowledge, make good use of, and perhaps even valorize this necessary condition of human life. Institutions will be built up around it, and they will begin with the expectations that trade will tend to take place, and that it will commonly be to mutual benefit.

A command economy can do no such thing. On the contrary, it is embroiled in a complex relationship with trade. It will have to tolerate some small-scale trade, as the machinery of repression only rarely grinds so finely, and as illicit trade would appear necessary to make the whole system work in the first place. But where repression works best is at suppressing the largest and most legible types of trade—the trades that involve great distances, great quantities, long stretches of time, extensive public notice, third parties whose trust cannot be assured, or complex assignments of property rights that require specialized knowledge for their adjudication. Trade of various types will fare variously under a communism, with the most legible actions being the most effectively suppressed. Barter, however, never fully goes away. At least given current and historically existing technologies, it's just too easy to conceal.

This qualitative difference between the command and market approaches is most clearly visible when whole societies act, or profess to act, only according to one principle or the other. But we should not infer that it does not apply to mixed economies as well. Even in the midst of a generally market-oriented society, to the extent that one finds attempts to command in favor of a particular goal, and to the extent that these attempts interfere with the desires of individual actors, strategies of evasion will arise. These strategies will aim at trade to mutual individual benefit despite laws to the contrary. They may take the form of black markets, informal barter economies, substituted goods, and even networks of organized violence outside the state. Associated losses will be less visible and smaller than in a wholly command economy, but they will be no less real.

A communist might be tempted to argue against this line of reasoning that a pure communism has never yet been achieved—and if it were, barter would not be needed. Since we have argued elsewhere that a pure capitalism has never yet been achieved, it would be uncharitable not to grant this objection. Under a pure communism, our interlocutor might continue, barter would not be *desired*, as no one would ever think of goods as particularly their own. And, he might continue, barter might not even be seen as advantageous; in a communism, the notion of a *personal* advantage is antiquated.

An interventionist who was not a communist might even argue likewise: in a better world, no strategies of evasion directed against the interventionist policy would be desired, for all would come to see it as being in their own best interests, rightly understood. Interventionism will work in any particular area at all, if only we achieve the proper inner consciousness of the participants.

Yet this would seem to stipulate into existence one of two things— either a material plenty beyond all wants that we could imagine, which would suffice to remove all thoughts of personal advantage; or a sheer imperviousness to material interest. In the case of interventionism, this imperviousness to material interest must be startlingly ad hoc as well: It must work in one area, that which the legislator has dictated, while not working in others, for which the natural impulse to trade must be counted on to function as normal. Interventionism would seem to demand not merely a New Socialist Man, but a New Socialist Man who can turn on a dime, at least if full compliance is expected.

These are thin reeds on which to rest a social system. Note that we would *not* achieve a disinterest in barter and other elementary forms of property-like behavior through perfect material equality alone: Tastes and affinities are always to some degree diverse, and, in a population with perfectly equal holdings of all goods, various consumers' diverse tastes would provide incentives to barter. At least some inequality would soon result. As G.A. Cohen notably lamented, "[T]hose who love work will, ceteris paribus, relish their lives more than those who hate work do." Ordinarily, one might be tempted to call this disposition a virtue; Cohen viewed it as a problem, and one that he feared his socialism could not fix.[4] But perhaps the desire for an ever-greater equalization of happiness is not one that can be adequately addressed by means of changes in our political or social institutions. Some aspects of happiness, or its opposite, may always lie with the individual.

As interesting as it may be in theory, communism in practice is passé. Even arguing against it may seem a wasted effort, although I hope it has not been one. The reason for arguing against communism has not been the fear of a communist resurgence. It has been, rather, to show how our political theories can fail to account for elementary and pervasive aspects of human behavior, such as trade. And even when they fail in this manner, they can seem highly persuasive anyway.

It may be urged that mixed economies *do* make room for trade, and that, as a result, they suffer none of the hypocrisies of communism in practice. This may be so. An adequate critique of a mixed economic system is a much more complex affair than an adequate critique of theoretical communism.

In part, that's because the latter never makes contact with reality. But, as mentioned above, private strategies of evasion are also less legible in a mixed economy. Both free institutions and greater personal wealth enable noncompliers to act as they wish in comparative privacy. Avenues of escape from bad policy may not look especially like a universal black market, or like desperate barter. They may not be apparent at all, and, in the last resort, it is to the interventionists' advantage to deny that they are even occurring. Things run more smoothly that way, or at least they appear to.

It is often claimed, and correctly, that poverty is the natural state of mankind. At least in Rousseau's chronological sense, the state of nature is indeed one of abject material poverty. But this is not to say that poverty is good, or virtuous, or that it is a thing that we all deserve. It is emphatically none of these, and Rousseau knew not whereof he spoke. Few if any will agree with Rousseau after understanding what the doctrine of the noble savage truly demands of them. And while we are far from ascertaining any clear reason why a state should exist, we may still explain why society should exist, with or without a state: Against Rousseau, society's purpose is to alleviate the dearth of the unimproved world through coordinated action. It is well and good that it does so.

Trade is one of the key social methods of alleviating dearth. *Labor* is another. Find wealth, and there you will typically find an artifice, a man-made thing, a thing that is beyond brute nature. Find wealth, and there you will find an effort, a stored-up supply of work, and an ingenuity of some sort. And even if labor is not *sufficient* to create value, it is in almost all cases *necessary*.

Yet seldom in recent economic history has very much of value been the effort of just one person. Much more often nowadays it has been a coordinated and cooperative effort that has produced whatever form of wealth or ease you may be considering. Along the way, many different people have performed many different types of work, and much trade has taken place unknown to the final consumer. Without such trade, the final good could never exist. Even the humble pencil, as Leonard Read famously observed, was the coordinated work of a production chain that spanned the entire globe and required the knowledge and effort of dozens of specialists. None of them could have made the pencil alone.

Pencils—and so much else—are social products. Given that nature provides us with so little, it is a testament to our sociability that so many of us nowadays have so much. Far from atomizing the individual, markets have brought us all to cooperate, and the product of that cooperation can be seen in every light bulb, in every pencil, in every computer and house. Tracing exactly how we get from the unimproved world to the one that most of us are accustomed to would be a hopelessly long undertaking, well beyond the powers of any one author. But a sketch of some general principles may explain what could otherwise seem like an unfounded assertion.

One too-simple way to think about production's place in the political world is often found in state-of-nature arguments about property:

An individual clears some land, builds a house, and plants crops. All of which—crops, house, and land—are asserted now to belong to him, to the exclusion of all others, for a variety of reasons. Locke argued that the mixing of labor with the land meant that by right the land had to attach to him as well. One may deny this argument, but only with the unsavory (though not necessarily insuperable) consequence of robbing a laborer of his labor value. Hume, however, argued that property existed because it was advantageous to mankind, in that it set up incentives that encouraged gainful activity and punished idleness and waste. I believe Hume's approach is considerably stronger as a description of property's role in civil society. As Hume argued, insisting that individuals be incentivized through property ownership is the rule that on the whole produces an industrious society. The rule is justified with reference to its consequences, which are on the whole, if not uniformly, good.

By contrast, allowing some individuals to appropriate the product of others' labor threatens to incentivize a caste society, one in which workers toil for the sake of the privileged few who merely appropriate. This, recall, was Oppenheimer's critique of the state, and it has often been a telling one.

And yet the relevant incentives in a property-holding society are hardly cabined by the boundaries of a parcel of initially appropriated land, one that must be defended from bandits. We are not subsistence farmers, and the story of trade must yet be told, at least in outline, so that we may incorporate it into an account of the good society.

Individuals who provide solely for themselves and their families, without the benefit of commerce or industrialization, will still routinely face dearth. They may have every good reason to enter into a Lockean social contract, but material culture in this mode of production will nonetheless be scanty; a poor harvest will mean famine, and diseases will seldom have any effective cures. Whenever warriors are a plausible threat, few dare spend their surplus on research, and among this society's other woes, knowledge will not advance.

Meanwhile, the warriors always threaten, and they must be defended against, which costs resources yet again, resources that individuals of good will may or may not have. Any particular state may or may not have arisen from a domesticated warrior class; any particular state may or may not have arisen from the attempts to repel someone else's *un*domesticated warrior class. The genealogy of the state matters little when compared to the enormous social losses at hand.

Now, to the individual warrior, being a warrior may seem great. It comes with all sorts of comforts, and not the least of them is the chance to look down upon the peasants, and to think oneself a higher order of being. To the ancient historians, this in itself seems to have been part of what government was for.

Yet every warrior brings on the tip of his spear at least *three* forms of loss: First, the efforts that workers devote to self-defense could have been spent on production, if the warriors had not presented their threat. Second, a warrior's efforts spent in warring could *also* have been spent on production. This again would have meant more goods in total. What the warrior "produces" is no production at all; it is merely a transfer. Viewed impartially, the warriors waste *everyone's* time, including their own.

And that brings us to the third and most serious form of loss from conquest: When one is reasonably confident that no warriors will arrive, and that one will be secure in reaping the fruits of one's labor, the simple, sole-producer mode of production begins to undergo a remarkable transformation. Individuals begin to specialize in what they are good at, and they then exchange their surplus product for things that they are not so good at producing. This is in accordance with what economists call *comparative advantage*. The benefits of comparative advantage created modern material culture, and they can't be had to anything like the same degree when the threat of plunder exists.

Even today, in a society that depends on comparative advantage, it seems clear that many do not understand this principle, and that its implications for political theory have been all too little appreciated. It can hardly be re-explained often enough, so I will do so here. Consider the following example:

> Suppose a lawyer can earn $100 per hour doing legal work, and can type 100 words per minute. Suppose secretaries can be hired for $10 per hour, but they can only type 50 words per minute. The lawyer has an absolute advantage in typing. Should he hire the secretary anyway? Yes, because the secretary has a relative advantage in typing. If he does his own typing, he saves $20 per hour (from two secretaries), but loses $100 in legal work, for a net loss of $80. He is better off hiring two secretaries to do what he could do in an hour. The same principle applies to trade between countries.[5]

Note that it's better for the lawyer to hire *two* secretaries, *even though he's better than they are at everything, including typing*. Why? Because if he doesn't, then (1) the lawyer doesn't produce as much lawyering and (2) the secretaries won't produce anything at all. Hiring them means that there is less superfast lawyerly typing in the world. But it also means that there is more overall value.

The workers' talents may be distributed in any way at all, and as long as they are somewhat unequal, two people will always produce more through coordination and trade. And even if they begin as equals, if they agree to specialize anyway, each person will soon probably get better at the task he or she has chosen, and gains from trade will emerge. All that needs to be done to incentivize this sort of coordination is to allow the participants a share of the surplus. If the transaction costs aren't too high, people will tend to take advantage of this opportunity, and their holdings will increase on the margin.

But this opportunity to specialize and trade will often be thwarted by those who see that they can also increase their holdings by deploying the political means. Because the political means interferes with specialization and trade, warriors don't simply waste everyone's production time—they also wreck our chances at pursuing comparative advantage. Yet again, the economic means and the political means are locked in a conflict with one another. But this time, gains from trade are at stake; they will either be captured by the industrious, or they will fail entirely to materialize.

The question we must then ask ourselves is simple: How much warrior-like activity must there be in our society? It is not simply communism that inhibits the pursuit of comparative advantage. All warrior-like activity does it to some extent, including the tranquil, domesticated forms that we see in the modern mixed economy. These too stand to inhibit the pursuit of comparative advantage, and a very simple fact should give us pause about endorsing such seemingly innocuous measures as a new tax or a new regulation. We do not yet know the limits of our own growth under the strategy of pursuing gains from trade, and it could be the case that we are even now all too inhibited.

That's because comparative advantage scales incredibly well. From two individuals, all the way up to the entire world considered as a working whole, comparative advantage goes right on re-ordering our activities toward the production of more and more output—at least, as long as individuals are incentivized toward it, and as long as institutions exist to facilitate rather than thwart it. Comparative advantage, and not Locke's simple, homesteaded property, is *the* reason why we are incomparably rich as compared to hunter-gatherers—even though, were we each left to our own devices, we might not even have food, shelter, or clothing.

Comparative advantage also works for nations, which is why free trade between them is a good idea. Comparative advantage works regardless of race or gender, which places it squarely in conflict with ideologies that take these as relevant characteristics for purposes of politics. Suitably qualified, comparative advantage works between and within firms and families, and indeed anywhere that tasks exist to be done.

I believe that a good society is at least in part one that facilitates gains from trade while inhibiting those things that would interfere with them. A good society is also one that draws people into the process of discovering comparative advantages—by rewarding them with a share of the surplus. A good society should therefore be suspicious of the strategy of unrequited transfer rather than proud of it. A good society should regard this strategy as a last resort, one that is constantly called into doubt.

Moreover, any system that presupposes that a bounty exists to be distributed is, to the best of our knowledge, implicitly predicated on the existence and good function of certain institutions, namely those that facilitate the pursuit of gains from trade: To claim that the key problem facing a given society is the achievement of the proper resource distribution is implicitly to claim that wealth creation has already and abundantly taken place, that it is ongoing without any impediment, and that the proposed distribution process will do nothing to significantly disturb it.

This, though, assumes a great deal. Under capitalism, wealth creation does not come at the moment of production, but rather at the moment of *trade*, that is, only at the moment that it reaches the consumer. What we may be tempted to think of as the capitalist mode of production, via factories and specialization of labor, would in fact amount to a colossal waste if trade were impossible or nonexistent.

It may not be clear why this is the case, so let's explain in some greater detail.

In economics, the law of declining marginal utility holds that, outside some rare and perhaps only theoretical circumstances, the second unit of a good is less useful than the first, and the third unit is less useful than the second, and so on. Often, additional units of a good are actually harmful to the possessor, particularly when we consider the costs of storage, maintenance, and the like. Over time, these so-called goods can even be ruinous.

Let us consider an example. A bread manufacturer may, like the rest of us, enjoy the product of his factory. But he certainly does not need, and he could not possibly eat, the hundredth loaf to come out of the industrial-sized oven in the course of a day. And still less could he possibly eat the thousandth. And then the bread starts to go bad, and his business is ruined. Why on earth would anyone ever make so much at once? Running a bread factory would be a symptom of insanity without the prospect of trade. It would amount to a gross waste of resources.

What we typically think of as the capitalist mode of production—that is, labor specialization and deep investment in production technology—would constitute a horrid destruction of wealth, if trade in consumer goods did not follow immediately afterward. It is only with trade that the hundredth or the thousandth loaf of bread—useless as such to the breadmaker—becomes the *first* loaf in a hundred or a thousand different households. Only trade can confer on a loaf of bread the status of being the first one consumed by my family this week, and with that status, it takes on considerable value.

Trade is what transforms these loaves from useless waste into literally our daily bread. Indeed, capitalist production may almost be defined as *that type of production which depends for its success on the prospect of large-scale trade*. The capitalist means of production, then, is not really the factory system. The capitalist means of production is the *market*, where goods that are of low marginal utility to one person become goods of high marginal utility to another. Seizing the factories for the benefit of the proletariat, as Marxists might do, does *not* seize the capitalist means of production. The capitalist means of production, which is trade, cannot be seized at all. It can only be permitted, or not, in thousands of different instances.

A Marxist might object as follows. Why not seize the factories—call them what you like—and distribute their products directly to the needy? Why argue about which aspect of economic activity is really the "means" of production? There are people who need bread right now, and there is a factory to supply it. Let us not be stopped by a quibble over nomenclature.

In the near term, the plan would probably work quite well: People who formerly didn't have bread would get it, and they would be better off. But what we have seen of the production and distribution of consumer goods under communism gives no reason for optimism beyond the first few moments of satisfaction. Moreover, distribution according to claimed need faces severe problems in efficiently allocating raw materials and production goods, as I have discussed elsewhere. The so-called socialist calculation problem and the experience of communism in practice, both suggest that any short-term success that might be had at feeding the poor by seizing a factory will come at a terrible cost later on, when supplying, maintaining, and upgrading the factory efficiently becomes important.

The capitalist system is able to attend to these long-term problems, while the aging factories of the late Soviet Union attest to a basic Marxist error: The factories never were the means of production. The means of production lay in the act of exchange for increased marginal utility. The true means of production had not been seized but rather suppressed, and derided, and little attended to, under the belief that factories themselves were all that were necessary. But it was trade that produced value, and it is trade that animates the good society even today. If I were forced to say what society—and not a government—is for, it would be this—the exchange of goods, services, ideas, and even affections, in a widening, virtuous circle.

So far, humanity would appear to have vacillated between two principles of social organization: The first endows a political class with all or substantially all goods that others have produced, or at least endows the political class with the means to seize them at will. Such a society yields little or no surplus value, and most within it experience grinding misery. Sad to say, this is the social system that far too much of western political theory has always endorsed. It has been the purpose of this book to suggest that our theories have failed us, and that we can and must do better.

The second social system is the one that results when all enjoy a measure of liberty to acquire, improve on, trade, and consume the products of their own labor, without fear that a political class will intervene. We do not yet know what our society would look like if this organizing principle were carried out consistently. Indeed, we do not yet know exactly how to implement such an expansive liberty. But we may know enough, based on hard-won experience, to determine the direction that we ought to move.

NOTES

1. Scott, *Two Cheers for Anarchism*, p. 47.
2. Pearson and Tudor, *North Korea Confidential*.
3. Gintis, "The Evolution of Private Property," available at http://www.umass.edu/preferen/gintis/Evolution%20of%20Private%20Property.pdf. Of course, merely because animals behave in a certain way, it does not follow that humans ought to do likewise.
4. Cohen, *Why Not Socialism?*, p. 21.
5. Foldvary, *The Science of Economics*. San Diego: Cognella Academic Publishers, 2010. ch 18, http://www.foldvary.net/sciecs/ch18.html.

BIBLIOGRAPHY

Amar, Akhil Reed. 2005. *America's Constitution: A Biography*. New York: Random House.

Aristotle. 1987. *Nichomachean Ethics*. Trans. David Ross. New York: Oxford University Press.

———. 1995. *Politics*. Trans. Ernest Barker, rev. R. F. Stalley. New York: Oxford University Press.

Ashcraft, Richard. 1989. *Revolutionary Politics and Locke's Two Treatises of Government*. Princeton: Princeton University Press.

Augustine of Hippo. 1887. *City of God*. Trans. Marcus Dods. *From Nicene and Post-Nicene Fathers, First Series*, Vol. 2. Ed. Philip Schaff. Buffalo: Christian Literature Publishing Co. Revised and edited for New Advent by Kevin Knight. http://www.newadvent.org/fathers/120101.htm

Bakunin, Mikhail. 1972. *Bakunin on Anarchy*. Ed. Sam Dolgoff. New York: Alfred A. Knopf.

Barnett, William, and Walter Block. 2007. Coase and Van Zandt on lighthouses. *Public Finance Review* 35: 710–733.

Bellamy, Edward. 2000 [1878]. *Looking Backward: From 2000 to 1887*. Bedfort: Applewood Books.

Bentham, Jeremy. 1838–1843. *The Works of Jeremy Bentham, Publish`ed Under the Superintendence of His Executor, John Bowring*, 11 vols, Vol. 3. Edinburgh: William Tait. http://oll.libertyfund.org/titles/1922#Bentham_0872-03_284

———. 1907. *An Introduction to the Principles of Morals and Legislation*. Oxford: Clarendon Press.

Berman, Harold J. 1983. *Law and Revolution: The Formation of the Western Legal Tradition*. Cambridge, MA: Harvard University Press.

© The Author(s) 2017
J. Kuznicki, *Technology and the End of Authority*,
DOI 10.1007/978-3-319-48692-5

Block, Walter. 2004. Libertarianism, Positive Obligations and Property Abandonment: Children's Rights. *International Journal of Social Economics* 31(3): 275–286.

Bloom, Allan. 1968. *The Republic of Plato, Translated, with Notes, an Interpretive Essay, and a New Introduction.* New York: Basic Books.

Boettke, Peter. 2000. *Socialism and the Market: The Socialist Calculation Debate Revisited.* New York: Routledge Library of 20th Century Economics.

Bossuet, Jacques-Bénigne. 1681. *Discours sur l'histoire universelle,* 8. http://www.samizdat.qc.ca/cosmos/sc_soc/histoire/hist_med/hist_universel.pdf. All translations by the author.

———. 1709. *La Politique tirée des propres paroles de l'Écriture sainte.* Paris: Pierre Cox.

Boyle, Katherine. 2013. Yes, Kickstarter Raises More Money for Artists Than the NEA. Here's Why That's Not Really Surprising. *Washington Post,* July 7. http://www.washingtonpost.com/blogs/wonkblog/wp/2013/07/07/yes-kickstarter-raises-more-money-for-artists-than-the-nea-heres-why-thats-not-really-surprising/

Brennan, Jason. 2014. *Why Not Capitalism?* New York: Routledge Press.

Buchanan, James M. 2003. Public Choice: Politics Without Romance. *Policy.* Available at http://grad.mercatus.org/sites/default/files/Buchanan-%20Politics%20without%20Romance.pdf

Campanella, Tommaso. 1981. *The City of the Sun: A Poetical Dialogue.* Berkeley: University of California Press.

Cassirer, Ernst. 1946. *The Myth of the State.* New Haven: Yale University Press.

Chodorow, Stanley. 1972. *Christian Political Theory and Church Politics in the Mid-Twelfth Century.* Berkeley: University of California Press.

Church, Jeffrey. 2010. The Freedom of Desire: Hegel's Response to Rousseau on the Problem of Civil Society. *American Journal of Political Science* 54(1): 125–139.

Cicero. 1841–42. *The Political Works of Marcus Tullius Cicero: Comprising His Treatise on the Commonwealth; and His Treatise on the Laws. Translated from the Original, with Dissertations and Notes in Two Volumes.* By Francis Barham, Esq. London: Edmund Spettigue. http://oll.libertyfund.org/title/546/83295. Accessed Feb 18 2014.

———. 1931. *De Finibus.* Trans. H. Harris Rackham. Cambridge, MA: Loeb Classical Library, Harvard University Press.

Coase, Ronald H. 1974. The Lighthouse in Economics. *Journal of Law and Economics* 17(2): 357–376.

Cohen, G.A. 2009. *Why Not Socialism?* Princeton: Princeton University Press.

Cohen, Joshua. 2010. *Rousseau: A Free Community of Equals.* Oxford: Oxford University Press.

Conway, David. 1995. *Classical Liberalism: The Unvanquished Ideal.* New York: St. Martin's Press.

Corn, David. 2012. SECRET VIDEO: Romney Tells Millionaire Donors What He REALLY Thinks of Obama Voters. *Mother Jones*, September 17. http://www.motherjones.com/politics/2012/09/secret-video-romney-private-fundraiser

Darnton, Robert. 1985. Readers Respond to Rousseau: The Fabrication of Romantic Sensitivity. In *The Great Cat Massacre and Other Episodes in French Cultural History*. New York: Vintage Books.

———. 1996. *The Forbidden Best-Sellers of Pre-Revolutionary France*. New York: W.W. Norton.

de Jasay, Anthony. 1998. *The State*. Indianapolis: Liberty Fund.

———. 2010. *Political Philosophy, Clearly: Essays on Freedom and Fairness, Property and Equalities*. Indianapolis: Liberty Fund.

de Soto, Hernando. 2000. *The Mystery of Capital: Why Capitalism Succeeds in the West and Fails Everywhere Else*. New York: Basic Books.

de Tocqueville, Alexis. 2012. *Democracy in America*. Trans. James T. Schleifer, ed. Eduardo Nolla. Indianapolis: Liberty Fund.

Diogenes Laertius. 1853. *The Lives and Opinions of Eminent Philosophers*. Trans. C.D. Yonge. London: Henry G. Bohn. Available at http://classicpersuasion.org/pw/diogenes/dldiogenes.htm

Dodson, Kevin E. 1997. Autonomy and Authority in Kant's Rechtslehre. *Political Theory* 25(1): 93–111.

Filmer, Robert. 1680. *Patriarcha; of the Natural Power of Kings*. London: Richard Chiswell. http://oll.libertyfund.org/titles/221

Fiorentini, Gianluca, and Sam Peltzman, eds. 1995. *The Economics of Organised Crime*. Cambridge: Cambridge University Press.

Foldvary, Fred E. 2010. *The Science of Economics*. San Diego: Cognella Academic Publishers.

Friedman, David. 1989. *The Machinery of Freedom: Guide to a Radical Capitalism*. 2nd ed. Chicago: Open Court Publishing. Available at http://daviddfriedman.com/The_Machinery_of_Freedom_.pdf

Gibbon, Edward. 1906. *The History of the Decline and Fall of the Roman Empire*. Ed. J.B. Bury with an Introduction by W.E.H. Lecky. New York: Fred de Fau and Co. Available at http://oll.libertyfund.org/titles/1365#Gibbon_0214-01_346

Gintis, Herbert. 2007. The Evolution of Private Property. *Journal of Economic Behavior and Organization*. Available at http://www.umass.edu/preferen/gintis/Evolution%20of%20Private%20Property.pdf

Gratian. 1993. *Treatise on Laws*. Trans. Augustine Thompson; with the Ordinary Gloss, Trans. James Gordley, and an Introduction by Katherine Christensen. Washington, DC: Catholic University of America Press, Studies in Medieval and Early Modern Canon Law vol. 2.

Hansen, Mogens Herman. 1989. *Was Athens a Democracy?: Popular Rule, Liberty, and Equality in Ancient and Modern Political Thought*, Historisk-filosofiske meddelelser 59. Copenhagen: Royal Danish Academy of Sciences and Letters.

Hayek, F.A. 1952. *The Counter-Revolution of Science: Studies on the Abuse of Reason*. New York: The Free Press.

———. 1976. *Law, Legislation, and Liberty, Vol II: The Mirage of Social Justice*. Chicago: University of Chicago Press.

Headley, John M. 1988. On the Rearming of Heaven: The Machiavellism of Tommaso Campanella. *Journal of the History of Ideas* 49(3): 387–404.

Hegel, G.W.F. 1977. *Phenomenology of Spirit*. Trans. A. V. Miller. Oxford: Oxford University Press.

———. *Philosophy of Right*. https://www.marxists.org/reference/archive/hegel/works/pr/prstate.htm

Herodotus. 1954. *Histories*. Trans. Aubrey de Sélincourt. New York: Penguin Classics.

Hexter, J.H. 1975. The Burden of Proof. *Times Literary Supplement* 24: 1250–1252.

Heyd, David. 1991. Hobbes on Capital Punishment. *History of Philosophy Quarterly* 8: 119–134.

Hobbes, Thomas. 1839–45. *The English Works of Thomas Hobbes of Malmesbury; Now First Collected and Edited by Sir William Molesworth. Bart*. London: Bohn. http://oll.libertyfund.org/titles/585#Hobbes_0051-03_621

———. *The Elements of Law Natural and Politic*. http://oregonstate.edu/instruct/phl302/texts/hobbes/elelaw.html

Höpfl, Harro, and Martyn P. Thompson. 1979. The History of Contract as a Motif in Political Thought. *The American Historical Review* 84(4): 920.

House, Edward Mandell. 1912. *Philip Dru: Administrator*. New York: B. W. Huebsch. http://oll.libertyfund.org/titles/1443#Kant_0332_338

Huemer, Michael. 2012. *The Problem of Political Authority: An Examination of the Right to Coerce and the Duty to Obey*. New York: Palgrave Macmillan.

Huffman, Carl A. 2015. Mathematics in Plato's Republic. In *The Frontiers of Ancient Science: Essays in Honor of Heinrich von Staden*, ed. Brooke Holmes and Klaus-Dietrich Fischer, 219–220. Boston: Walter de Gruyter.

Hulliung, Mark. 1974. Patriarchalism and Its Early Enemies. *Political Theory* 2(4): 410–419.

Hume, David. 1896. *A Treatise of Human Nature Reprinted from the Original Edition in Three Volumes and Edited, with an Analytical Index*, by L.A. Selby-Bigge. Oxford: Clarendon Press. http://oll.libertyfund.org/title/342/55127

———. 1985. *Essays Moral, Political, and Literary*. Indianapolis: Liberty Classics.

Huxley, Aldous. 2005. *Brave New World and Brave New World Revisited*. New York: Harper Perennial.

Irwin, William. 2015. *The Free Market Existentialist: Capitalism Without Consumerism*. Chichester: Wiley Blackwell.

Isakov, Leah, Andrew W. Lo, and Vahid Montazerhodjat. 2015. *Is the FDA Too Conservative or Too Aggressive?: A Bayesian Decision Analysis of Clinical Trial Design*. Available at SSRN: http://ssrn.com/abstract=2641547 or http://dx.doi.org/10.2139/ssrn.2641547

Kant, Immanuel. 1784. *What Is Enlightenment?* Available at http://www.columbia.edu/acis/ets/CCREAD/etscc/kant.html

———. 1886. *The Metaphysics of Ethics* [1796] by Immanuel Kant. Trans. J.W. Semple. Ed. with Introduction by Rev. Henry Calderwood, 3rd ed. Edinburgh: T. & T. Clark.

———. 1889. *Kant's Critique of Practical Reason and Other Works on the Theory of Ethics*. Trans. Thomas Kingsmill Abbott, B.D., Fellow and Tutor of Trinity College, Dublin, 4th revised ed. London: Kongmans, Green and Co. http://oll.libertyfund.org/titles/360

———. 1891. Idea for a Universal History from a Cosmopolitan Point of View. In *Kant's Principles of Politics, Including His Essay on Perpetual Peace. A Contribution to Political Science*. Trans. and Ed. W. Hastie. Edinburgh: Clark. http://oll.libertyfund.org/titles/358#Kant_0056_36

———. *Critique of Pure Reason*, 526 ff. Available at http://oll.libertyfund.org/titles/1442#Kant_0330_1020

Kołakowski, Leszek. 1978. *Main Currents of Marxism*. Oxford: Oxford University Press.

Kuznicki, Jason. 2000. *Reasonable Souls: Jews, Christians, and the Catholic Origins of the Enlightenment in Claude Fleury's Moeurs*. MA diss, The Ohio State University, Columbus.

LeBar, Mark. forthcoming. The Virtue of Justice, Revisited. In *The Handbook of Virtue Ethics*, ed. Stan Van Hooft. Cambridge: Cambridge University Press.

Lévi-Strauss, Claude. *Jean-Jacques Rousseau, Fondateur des sciences de l'homme*. Address at the Université Ouvrière de Genève on the 250th Anniversary of Rousseau's Birth. Available at http://www.espace-rousseau.ch/f/textes/levistrauss1962.pdf

Levy, Jacob T. 2014. *There's No Such Thing as Ideal Theory*. Available at SSRN: http://ssrn.com/abstract=2420125

———. 2015. *Rationalism, Pluralism, and Freedom*. Oxford: Oxford University Press.

———. *On Taking Politics Less Seriously*. https://www.youtube.com/watch?v=cyRUJtfRfK8, at 30:04.

Livy. 1857. *History of Rome by Titus Livius, the First Eight Books. Literally Translated, with Notes and Iillustrations*, 1.1.Trans. D. Spillan. London: Henry G. Bohn.

Locke, John. 1988. *Two Treatises of Government*. Ed. Peter Laslett. Cambridge: Cambridge University Press.

———. 1993. *John Locke: Political Writings*. Ed. David Wootton. New York: Mentor.

Lomasky, Loren E. 2005. Libertarianism at Twin Harvard. *Social Philosophy and Policy* 22(1): 178–199.

Long, Roderick T. 1998. Civil Society in Ancient Greece: The Case of Athens. Given at the Liberty Fund Conference on Civil Society, Arlington, 29 May 1998. At http://www.praxeology.net/civsoc.htm

Lucian of Samosata. 1905. A Slip of the Tongue in Salutation. In *The Works of Lucian of Samosata*. Trans. H. W. Fowler and F. G. Fowler. Oxford: The Clarendon Press.

Machiavelli, Nicolò. 1992. *The Prince*. Trans. Robert M. Adams, 2nd ed. New York: Norton Critical Edition.

———. 2003. *Discourses*. Trans. Leslie J. Walker, S.J. New York: Penguin Books.

Marx, Karl. *Capital* [1867]. Trans. Samuel Moore and Edward Aveling [1887]. Available at https://www.marxists.org/archive/marx/works/1867-c1/ch15.htm

———. A Contribution to the Critique of Hegel's Philosophy of Right. Ed. Andy Blunden and Matthew Carmody. Available at https://www.marxists.org/archive/marx/works/1843/critique-hpr/intro.htm

Marx, Karl, and Friedrich Engels. 1969. *Marx/Engels Selected Works*. Moscow: Progress Publishers.

McAlpin, Mary K. 2010. Innocence of Experience: Rousseau on Puberty in the State of Civilization. *Journal of the History of Ideas* 71(2): 241–261.

Menake, George T. 1982. Research Note and Query on the Dating of Locke's Two Treatises. *Political Theory* 10(4): 609–612.

Milton, John. 1650. *The Tenure of Kings and Magistrates*. London: Matthew Simmons.

Milton, J.R. 1995. Dating Locke's Second Treatise. *History of Political Thought* 16(3): 356–390.

Milton, Philip. 2000. John Locke and the Rye House Plot. *The Historical Journal* 43(3): 660.

Moore, R.I. 1990. *The Formation of a Persecuting Society: Power and Deviance in Western Europe, 950–1250*. Cambridge, MA: Blackwell Publishers.

Morley, Neville. 2000. Trajan's Engines. *Greece and Rome* (Second Series) 47: 197–210.

Nozick, Robert. 1974. *Anarchy, State, and Utopia*. New York: Basic Books.

Olson, Mancur. 2000. *Power and Prosperity: Outgrowing Communist and Capitalist Dictatorships*. New York: Basic Books.

Oppenheimer, Franz. 1914. *The State: Its History and Development Viewed Sociologically*. Trans. John M. Gitterman. Indianapolis: Bobbs-Merrill. Repr. 2012, Forgotten Books.

Ostwald, Martin. 1990. *Nomos and Phusis in Antiphon's Peri Alétheias.* UC Berkeley: Department of Classics, UCB. Retrieved from: http://escholarship.org/uc/item/7kg1w5zm

Paine, Thomas. 1894. *The Writings of Thomas Paine, Collected and Edited by Moncure Daniel Conway.* New York: G.P. Putnam's Sons. http://oll.libertyfund.org/title/344/17372

Palmer, Tom G. 2009. Classical Liberalism, Marxism, and the Conflict of Classes: The Classical Liberal Theory of Class Analysis. In *Realizing Freedom: Libertarian Theory, History, and Practice.* Washington, DC: The Cato Institute.

Parfit, Derek. 1984. *Reasons and Persons.* Oxford: Oxford University Press.

Pearson, James, and Daniel Tudor. 2015. *North Korea Confidential: Private Markets, Fashion Trends, Prison Camps, Dissenters and Defectors.* North Clarendon: Tuttle Publishing.

Plato. 1961. *Plato: The Collected Dialogues.* Ed. Edith Hamilton, Bollingen Series LXXI. Princeton: Princeton University Press.

Plutarch. 1919. *Lives.* Trans. Bernadotte Perrin. Cambridge, MA: Harvard University Press.

———. 1936. *Moralia.* On the Glory of the Athenians. Cambridge, MA: Harvard University Press.

Pocock, J.G.A. 1975. *The Machiavellian Moment: Florentine Political Thought and the Atlantic Republican Tradition.* Princeton: Princeton University Press.

Political Fragments of Archytas and Other Ancient Pythagoreans. 1822. Trans. Thomas Taylor, 106–07. Available at http://en.wikisource.org/wiki/Political_fragments_of_Archytas_and_other_ancient_Pythagoreans/How_we_ought_to_conduct_ourselves_towards_our_kindred

Polybius. 1889. *Histories* 1.1. Trans. Evelyn S. Shuckburgh. New York: Macmillan. Reprint Bloomington 1962.

Popper, Karl. 2002. *The Logic of Scientific Discovery.* New York: Routledge.

———. 2003. *The Open Society and Its Enemies.* New York: Routledge Classics.

Rasmussen, Douglas, and Douglas J. Den Uyl. 2005. *Norms of Liberty: A Perfectionist Basis for Non-Perfectionist Politics.* University Park: Pennsylvania State University Press.

Rawls, John. 1999. *A Theory of Justice.* Cambridge, MA: Harvard University Press.

Rothbard, Murray. 1998. *The Ethics of Liberty.* New York: New York University Press.

Rousseau, Jean-Jacques. 1923. A Dissertation on the Origin and Foundation of the Inequality of Mankind. In *The Social Contract and Discourses.* Trans. with an Introduction by G. D. H. Cole. London/Toronto: J.M. Dent and Sons. http://oll.libertyfund.org/titles/638#Rousseau_0132_654

———. Constitutional Project for Corsica. http://www.constitution.org/jjr/corsica.htm

———. *Confessions*, book x. Trans. by the author. French text available at http://athena.unige.ch/athena/rousseau/confessions/rousseau_confessions_10.html

Sandefur, Timothy. 2004. Liberal Originalism: A Past for the Future. *Harvard Journal of Law & Public Policy* 27: 489.

Schaefer, Peter F., and Clayton Schaefer. 2014. *An Innovative Approach to Land Registration in the Developing World: Using Technology to Bypass the Bureaucracy*, Cato Policy Analysis 765. Washington, DC: The Cato Institute.

Schmitt, Carl. 1996. *The Concept of the Political*, Expanded edition. Chicago: University of Chicago Press.

Schwartz, Joel. 1985. *The Sexual Politics of Jean-Jacques Rousseau*. Chicago: University of Chicago Press.

Scott, James C. 2012. *Two Cheers for Anarchism*. Princeton: Princeton University Press.

Shklar, Judith N. 1969. *Men and Citizens: A Study of Rousseau's Social Theory*. New York: Cambridge University Press.

Skinner, Quentin. 1978. *The Foundations of Modern Political Thought*. Cambridge: Cambridge University Press.

———. 1989. The State. In *Political Innovation and Conceptual Change*, ed. Terence Ball, James Farr, and Russell L. Hanson. Cambridge: Cambridge University Press.

———. 2008. *Hobbes and Republican Liberty*. Cambridge: Cambridge University Press.

Smith, Adam. 1904. *An Inquiry into the Nature and Causes of the Wealth of Nations*. Ed. Edwin Cannan. London: Methuen & Co. Ltd.

Sommerville, J.P. 1982. From Suarez to Filmer: A Reappraisal. *The Historical Journal* 25(3): 525–540.

Spencer, Herbert. 1851. *Social Statics: Or, the Conditions Essential to Happiness Specified, and the First of Them Developed*. London: John Chapman. http://oll.libertyfund.org/titles/273#Spencer_0331_46

Stringham, Edward Peter. 2015. *Private Governance: Creating Order in Economic and Social Life*. Oxford: Oxford University Press.

Szabo, Nick. Formalizing and Securing Relationships on Public Networks. Available at http://szabo.best.vwh.net/formalize.html

Tabarrok, Alexander. 1998. The Private Provision of Public Goods Via Dominant Assurance Contracts. *Public Choice* 96: 345–362. http://mason.gmu.edu/~atabarro/PrivateProvision.pdf

The Federalist. 2001. Gideon edition, ed. George W. Carey and James McClellan. Indianapolis: Liberty Fund.

The Older Sophists: A Complete Translation by Several Hands of the Fragments in Die Fragmente der Vorsokratiker, Edited by Diels-Kranz. 1972. With a New

Edition of Antiphon and of Euthydemus. Rosamond Kent Sprague, ed., 220–221. Columbia: University of South Carolina Press.

Tolstoy, Leo. 2011. *The Kingdom of God Is Within You: Christianity Not as a Mystic Religion but as a New Theory of Life.* Trans. Constance Garnett. Amazon Digital Services LLC.

Tomasi, John. 2012. *Free Market Fairness.* Princeton: Princeton University Press.

Van Kley, Dale K. 2008. Civic Humanism in Clerical Garb: Gallican Memories of the Early Church and the Project of Primitivist Reform 1719–1791. *Past and Present* 200(1): 77–120.

Vanderbilt, Tom. 2008. The Traffic Guru. *Wilson Quarterly.* http://www.wilson-quarterly.com/article.cfm?AID=1234

Varro, Marcus Terentius. *De Re Rustica.* Available at http://penelope.uchicago.edu/Thayer/E/Roman/Texts/Varro/de_Re_Rustica/1*.html

Verhaegh, Marcus. 2004. Kant and Property Rights. *Journal of Libertarian Studies* 18(3): 11–32.

von Mises, Ludwig. 2013. *Epistemological Problems of Economics.* Trans. George Reisman. Indianapolis: Liberty Fund.

Weiner, Mark S. 2013. *The Rule of the Clan.* New York: Farrar, Straus and Giroux.

White, Mark D. 2009. *Adam Smith and Immanuel Kant: On Markets, Duties, and Moral Sentiments.* Prepared for the Association of Social Economics Meetings at the 2009 ASSA Annual Conference. Available at http://papers.ssrn.com/sol3/papers.cfm?abstract_id=1318605

White, Richard. 2010. *The Middle Ground: Indians, Empires, and Republics in the Great Lakes Region, 1650–1815.* Cambridge: Cambridge University Press.

White, Mark D. 2011. *Kantian Ethics and Economics: Autonomy, Dignity, and Character.* Stanford: Stanford University Press.

Zwolinski, Matt. 2015. Libertarianism and Pollution. In *The Routledge Companion to Environmental Ethics,* ed. Benjamin Hale and Andrew Light. New York: Routledge. Available in draft at http://papers.ssrn.com/sol3/papers.cfm?abstract_id=2443030

INDEX

Note: Page numbers followed by 'n' refers to notes.

© The Author(s) 2017

J. Kuznicki, *Technology and the End of Authority*,

DOI 10.1007/978-3-319-48692-5

Euthyprho, 23
Exclusion Crisis, 98
externalities, 227, 228

F
falsification, 136, 221–37
fascism, 188
feudalism, 175
fiat currency, 231
Filmer, Robert, 4, 60, 80, 82, 83,
 85n40, 141n3
Foucault, Michel, 196
France, 72, 84n14, 118, 119, 201,
 203
Frankfurt School, 241
French Revolution, 88, 118
Friedman, David, 107, 142n45

G
Gay New York (Chauncey, George),
 196
General Motors, 247
general will, 44, 117, 118, 131,
 151
Genoa, 117
Germanic realm (Hegel), 159
German Ideology, The (Marx, Karl),
 165, 181n23, 181n24
ghettos, 54
Gibbon, Edward, 81, 82, 86n61,
 195
Glaucon, 23
Glorious Revolution, 98
Gratian, 48–51, 53, 55, 56n20, 138,
 190
Great Chain of Being, 59, 75, 77,
 170
Great Fire of Rome, 42
Great Leap Forward, 155
Greece, 2, 23, 25, 27

Greek realm (Hegel, G.W.F), 158
*Groundwork for the Metaphysics of
 Morals* (Kant, Immanuel), 121

H
Harrington, James, 61, 84n14
Hayek, F.A., 4, 50, 137–9, 145n85,
 146n92, 196, 198n6
Hegel, G.W.F., 1, 44, 85n48, 128,
 147–63, 172, 180n1, 180n7–10,
 181n15, 181n19, 190, 198n7,
 200, 208n1
Hegelianism, 160, 161, 163
Herodotus, 9, 10, 29n4, 132, 133
Hierocles, 35
Hippo, 39
History of Rome (Livy), 62
Hobbes, Thomas, 5, 60, 77, 79–83,
 86n55, 88–97, 99, 114, 141n19,
 193
Holocaust, 155, 240
Höpfl, Harro, 89, 140n1
Horace, 75
humanism, 70, 84n25, 155
Hume, David, 65, 71, 84n11, 105,
 126, 142n43, 172, 182n30, 195,
 202, 208n3, 208n4, 214, 260
Hunger Games, The (Collins,
 Suzanne), 241
hunter-gatherers, 103, 262
Huxley, Aldous, 140, 146n97, 241

I
ideal act theory, 217, 218
individualism, 29, 89, 113, 120, 134,
 156, 159, 196, 201
industrialism, 175
Industrial Revolution, 164, 170
interventionism, 257, 258
Ion, 23

CPSIA information can be obtained
at www.ICGtesting.com
Printed in the USA
LVHW081618310319
612458LV00003B/10/P